MW01483941

ALSO BY DAVID WEITZNER

Connected Capitalism
Fifteen Paths

THINKING LIKE A HUMAN

THINKING LIKE A HUMAN

THE POWER OF YOUR MIND IN THE AGE OF AI

DAVID WEITZNER

Cover design by Jillian Rahn/Sourcebooks
Cover image © Mehau Kulyk/Science Photo Library/Getty Images
Internal design by Tara Jaggers/Sourcebooks

Published by Sourcebooks
P.O. Box 4410, Naperville, Illinois 60567-4410
(630) 961-3900
sourcebooks.com
Library of Congress Cataloging-in-Publication Data
Names: Weitzner, David, author.
Title: Thinking like a human : the power of your mind in the age of AI / David Weitzner.
Description: Naperville : Sourcebooks, 2025. | Includes bibliographical references. | Summary: "How the World Really Works meets Bill Hammack in Thinking Like a Human by Dr. David Weitzner; take a fascinating journey through AI: from its origins in history to its influences on our everyday lives and the unethical way it's wielded by big tech; through it all, Weitzner explores how creators everywhere practice "artful intelligence"-and shows us a hopeful vision for the future, in which we think like human beings in the age of AI"-- Provided by publisher.
Identifiers: LCCN 2024043300 (print) | LCCN 2024043301 (ebook) | (hardcover) | (epub) | (pdf)
Subjects: LCSH: Thought and thinking. | Artificial intelligence.
Classification: LCC BF441 .W449 2025 (print) | LCC BF441 (ebook) | DDC 153.4/2--dc23/eng/20241023
LC record available at https://lccn.loc.gov/2024043300
LC ebook record available at https://lccn.loc.gov/2024043301

Printed and bound in the United States of America.
MA 10 9 8 7 6 5 4 3 2 1

To my children, Moishe, Shaindy, and Leah, whose artfulness creates a world worth living in

Contents

Introduction

A TALE OF THREE SOUPS

For better or worse, revolution is always in the air.

In business and tech, the revolutionary aspiration is for breakthroughs that can catalyze an unexpected tomorrow. Few in these fields have the patience, or venture capitalist funding, to work for incremental outcomes. Nobody opens their wallets after being pitched a future only slightly better than today. So tech boosters take their cues from carnival hucksters and televangelists, flooding social media with passionate confessionals and messianic zeal. Like a few days into the initial release of ChatGPT, when online message boards barked that this incredible new tool was going to change *literally everything*. The world outside of Silicon Valley and the virtual chat rooms of Reddit had no idea what was brewing, as a profoundly different future arrived a little early in tech-forward spaces.

One tuned-in radical[1] was hungry for more than new toys to play with. Unable to imagine a satisfying dish to quickly put together, he turned to the tech of tomorrow, typing an inventory of all the food in the cupboards and refrigerator, and prompting a chatbot to derive a tasty recipe utilizing items on the list. In an instant, the algorithm came up with an unexpected delicacy: garbanzo bean and cheddar cheese–infused chicken soup. Having consumed the creation, the now satiated tech enthusiast ecstatically reported how blown away he was with the result. Who would have ever thought to make chicken soup?!? And why would a hungry soul return to the dark ages of self-directed meal construction when there's now an artificial intelligence (AI) ably fit for the task?

That moment of change was a few years back. Today, the generative AI future has arrived for almost everyone with a computer. So…are you more enlightened? Has the practice of outsourcing meal planning to AI been widely adopted? Have algorithms significantly improved your daily existence? In my circle, we're still waiting for the upending of worlds, knowing that we can access any number of different processes, algorithmic and other, to end up with an equally mediocre chicken soup. Like use the chatbot to recommend a recipe, rather than craft it. Or engage older technology, like a search engine, to easily find a foodie website and follow the human-constructed algorithm of steps. Lacking digital aid, maybe there's a recipe that has been passed down in the family, or a favorite to grab from a cookbook. If we're picky, we could narrow the algorithm behind our recipe search to a particular cultural style of chicken soup preparation that aligns with our tastes, or limit the ingredients to what is accessible.

If you're feeling the counterrevolutionary spirit, there is another possible fix to the soup-making dilemma that relies on truly outdated tech. For despite their merits, none of these algorithmic approaches are the way my beloved grandmother made her chicken soup, a dish of legend within our family and local community. She never followed a strict recipe. And she certainly didn't have access to an AI like ChatGPT to help her out. But she did have access to a wonderful assortment of cognitive tools that served her, and all those cooking before her, quite well in the quest to nurture a hungry crowd.

You see, my old-fashioned grandmother relied on an elaborate and holistic form of natural, artful intelligence. She would, for instance, taste the soup as it cooked, making all sorts of adjustments that felt right in the moment. Some days might have called for a spicier soup, while others a milder taste. There might be a time for *kneidlach* (matzo balls), a time for *lokshen* (noodles), or a time for *kreplach* (dumplings), depending on which Jewish holiday season we were in. I'm sure many of you have recollections of something, or someone, similar. There were likely no rules behind our grandparents' soups. Just a skill for transposing nurturing warmth into liquid goodness.

We Still Have Options

Society stands at the very start of an AI revolution. Yet already, too many of us peering into a near-term future are overcome with feelings of dread, not hope. We see a world substantively the same on quality-of-life metrics, made worse by tech deployed to destroy jobs and upend the ability of many to stay productive. But there is nothing necessary about any technological advancement. There is no reason to view corporate control of AI advancements as inevitable. Nor are we predestined to embrace whatever products Big Tech decides to sell us down the road. Future AI can look like whatever a researcher or engineer wants it to. And it is our choice, individually and collectively, to decide whether we use these new technological tools as cognitive aids or as crutches.

Like the artful intentions of my grandmother crafting her soup, this book is designed to dish out warmth and optimism. The possibilities for how human intelligence will interact with artificial intelligence are very much open-ended. Revolutions bring with them unique, possibly once-in-a-generation, opportunities to rethink the value of that which is being disrupted. Since ours is a revolution in intelligence, it gives us the chance to reimagine the very act of thinking. Imagining is critical to shaping the social and technological future differently than current trends. The good news is that the skills required to think like a human are latent within us. To unlock this power, we're going to engage in one extended thought experiment: What happens if we more frequently swap out algorithmic thinking for artful intelligence?

"More frequently" doesn't mean exclusively. If you've had a busy day, find that your brain is a little fuzzy, know that you are hungry but unsure what to make, and want to GPT a recipe…no problem. Use the tools modernity affords, while ensuring you still know how to create a meal plan using your five senses. The human story is intrinsically tied to the evolution of the tools and resources used to build new worlds. This book won't shy away from encouraging the use of digital supports that make us more creative, innovative, or effective. But the gadgets we embrace must always make us better embodied beings, not hold us back from reaching our natural potential.

Soup-making was the first message broadcast to the mainstream that corporate efforts are afoot designed to change not only how we think but also how we work, create, and live. There have been many more missives since then, each looking increasingly wondrous at the outset, often in direct proportion to how problematic it becomes after the initial feeling of novelty faded. Carefully reflect on what tech boosters are asking us to give up as our society normalizes accepting the superiority of AI and other forms of algorithmic intelligence. Drawing on artful intellectual resources can help us imagine viable alternatives to the AI paradigms being forced on an ill-informed and inadequately prepared population.

What do you give up when you follow the algorithm? If you had asked Nintendo's CEO Satoru Iwata, he would have said your very humanity. The year 2023 was brutal for the video game industry.[2] Facing rising interest rates, inflation, and the end of easy access to cash, CEOs at studios like Amazon Games, Ubisoft, Epic Games, and Niantic followed the algorithm taught in business schools: when in a financial crunch, improve efficiency. Map out all the value-adding components, like logistics, marketing, research and development, or customer care, that go into the creation of your product and see where you can reduce costs. Usually the biggest opportunity for improving "efficiency" is cutting large numbers of inefficient humans. It's nothing personal, of course, just the algorithm of good business.

But back in 2011, facing similar pressures to let people go, Iwata did not follow the algorithm. Instead, he led his company through a period of financial upheaval by thinking like a human. Iwata stood strong, explaining that seeking short-term efficiencies will have a human cost that the algorithm does not include in its calculus, like lower morale and increased fear in those remaining with the company.[3] Iwata was not prepared to ignore his humanity. Nor did he give retailers cause to lose faith in Nintendo. Rather than take steps that would quickly increase revenue, he slashed the price of the 3DS portable system to encourage shops to keep stocking the hardware.[4] Finally, in a move that would have no material impact on the company's overall finances but was loaded in human-to-human symbolic value, Iwata took a 50 percent pay cut on his salary for that difficult year. Nintendo emerged from the crisis as a powerhouse because of this care. Iwata offered a lesson for future innovators that we will pick up on. Novelty emerges when we break the algorithm—it creates the opportunity for our humanity to shine.

The Book's Algorithm

This book is divided into four parts. Part I homes in on the problems created by our increased dependence on algorithms. Along with exploring the history and evolution of algorithmic thinking, we will ask what has been gained, versus what gets increasingly lost, in the rise of the algorithm. Assessing trade-offs is especially pertinent to the current economic revolution, where AI-infused computer networks have started to communicate among themselves without human intervention, giving rise to "deep" machine learning and a business model that is more hustle than fair trade.

Algorithmic supremacists argue that widespread AI implementation is designed to free up our time, so that we can be more creative and productive in an idyllic future. But becoming more creative or productive doesn't just happen; we need the right cognitive tools. To that end, Part II will teach us *how* to think artfully. We'll do a deep dive into research on embodied cognition, intention, and free will. We'll learn ways to be more curious about how our senses, feelings, and emotions can take on problem-solving responsibilities.

Part III will share the exhilarating narratives of those who model artful intelligence in their creative lives. We will meet a rock star, a therapist, and an inventor, three individuals who are blazing an artful approach to innovation by going against conventional logic, looking at problems through a big-picture lens, and breaking the patterns that prevent paradigm-shifting progress. We will learn from their words and experiences how to incorporate the tools behind their successes into our lives.

Part IV will turn its focus to artful living, exploring how to bring this type of thinking to our day-to-day lives, bettering the ethics behind our technological innovations and the businesses that provide them, creating physical spaces to thrive in, and emphasizing upward mobility for all, not just the owners of algorithms.

Artful thinking and living reasserts human agency, and rebuilds trust, embodied human to embodied human. This is the path for reversing the confusion accompanying our increasing dependence on AI and algorithmic thinking. This is the way of hope.

PART I

ALGORITHMIC DEPENDENCE

1

—

What's in a Name?

HOW CAN I MISLEAD YOU TODAY?

Names have potency. Within a name are clues about the history, character, or essence of the named entity. In many cultures, parents name their children after those they loved or admired in the hopes of bringing the qualities of the eponym to a new generation. So much of our personal identity is tied to our names, whether they were given at birth or we acquire them for ourselves as we grow. By carrying a name, one takes on the responsibility to live up to the personal and social expectations it brings.

Corporate names are no different from individual names in this regard. For example, Nike, known as the brand behind athletic triumph, seems to be smartly named after the Greek goddess of victory. So it may surprise you to learn that at its founding, the organization was known as Blue Ribbon Sports.[1] It took seven years before Jeff Johnson, the company's earliest hire, woke up with "Nike" as a name for the company's first branded shoe line, responding to the pressure of a 9:00 a.m.

factory deadline for a label. Founder and CEO Phil Knight didn't like the name but due to the time crunch conceded, "I guess we'll go with the Nike thing for a while." And Western pop culture was forever changed.

Making accurate inferences from names is so important to our sense-making that misleading names stick with us as frustratingly incongruent. The top three for me are that German chocolate cake has nothing to do with the nation but one Samuel German, Baker's Chocolate is not so named because of the profession but after Walter Baker, and Main Street was never San Francisco's primary thoroughfare, instead memorializing Charles Main.

More recently, Air Canada named a chatbot to be an automated part of its customer service team, designed to bring efficiency to services that "did not require a human touch."[2] Yet when the algorithm told Jake Moffatt that he could pay full fare on a ticket purchase facilitated by the algorithm to fly to a funeral and then apply for a bereavement discount later, Air Canada refused to honor the deal. The company argued that the chatbot should actually be named as a separate legal entity from the airline, responsible for its own actions. I guess making things up while taking a customer's money is one of the services so embedded in Air Canada's business model that it indeed no longer required "a human touch" to be efficiently executed. Unfortunately for Air Canada's corporate team, a tribunal's ruling named the bot a representative of the company and an extension of its website, setting an important legal precedent on corporate responsibility for algorithms.

"Algorithms" are the finite, predetermined sequences of steps the rationally inclined follow to help solve a problem. Although in the modern age the moniker has been popularized by computer scientists, algorithms are not only used in technology and programming. Humans have long used this tool as a mode of mental processing, a state we have named algorithmic thinking, like following a recipe for chicken soup. Some prior individual went on a creative exploration in soup-making, was pleased with the result, and codified the process for future humans (and now, it seems, machines) to replicate. The step-by-step breakdown is the defining characteristic of an algorithm. These sequences create the order our brains need to facilitate the mental processing of information too complicated to synthesize instantaneously.

The word "algorithm" comes from a name, in tribute to Muhammad ibn Mūsā al-Khwārizmī, a Persian mathematician whose surname was Latinized to *Algorismus*. Bearing the influence of the Greek word *arithmos*, which means "number," *Algorismus* became "algorithm," denoting the type of decimal arithmetic al-Khwārizmī popularized, only acquiring its modern linguistic meaning in the nineteenth century.[3] Knowing that the term comes to honor a mathematician is insightful, as algorithms are the ordered steps, the if/then processes, one uses to solve a mathematical problem.

Al-Khwārizmī was employed in the "House of Wisdom" established by the Caliph Harun al-Rashid of Baghdad in the late eighth century, where thinkers worked on everything from translating to building scientific tools.[4] Al-Khwārizmī's lasting contribution to intellectual history is in the mathematical operations for solving equations, which laid the foundation for another type of math named in his honor: algebra. This label was drawn directly from the Arabic title of al-Khwārizmī's book *Al-Jabr*. Al-Khwārizmī composed the text in his native tongue around 820, which was later translated into Latin around 1145, and used as the principal math textbook in European universities until the sixteenth century.[5] We in the English-speaking world should be grateful that it was the Arabic title that stuck, as the English translators of the book went with the much wordier *The Compendious Book on Calculation by Completion and Balancing*. That just doesn't quite roll off the tongue as smoothly as "algebra."

We can find intellectual roots of AI extending even earlier than al-Khwārizmī, in four-thousand-year-old Babylonian clay tablets filled with cuneiform script.[6] Computer scientist Donald Knuth explains that these "are genuine algorithms...they represented each formula by a step-by-step list of rules for its evaluation, i.e., by an algorithm for computing that formula. In effect, they worked with a 'machine language' representation of formulas."[7] The efforts to build AI are not new or recent. Humans have been hard at work setting the stage over millennia.

Math as Philosophy

Nearly a thousand years after al-Khwārizmī, a seventeenth-century German philosopher, Gottfried Wilhelm Leibniz, would expand the algorithmic methodology, bringing society another step closer to the digital age. Not only was Leibniz obsessed with linking math to philosophy, but he also built rudimentary computational devices. Using the help of clockmakers, mechanics, artisans, and a butler, Leibniz's team constructed a hand-crank-operated tool with cylinders of polished brass and oaken handles that claimed to calculate addition, subtraction, multiplication, and division.[8] Through these efforts, Leibniz ultimately advanced what digital researcher Jonathan Gray calls "a kind of computational imaginary—reflecting on the analytical and generative possibilities of rendering the world computable."[9]

I was first turned on to Leibniz as an undergrad studying philosophy. One quirk that made Leibniz memorable was his preference for composing personalized, often confrontational, letters. Instead of writing one-size-fits-all generic tomes targeting everybody and nobody, Leibniz used the written word to spark dialogue with folks he disagreed with. I imagine that if he were alive today, Leibniz would be a prolific Tweeter. All these very social exchanges suggest that despite being a dedicated rationalist, he did not want to live only in his head. But consistent with his rationalist beliefs, Leibniz tried to turn every sensual experience into a purely algorithmic mental exercise.

I have a vivid recollection of five-year-old me watching an early episode of *Sesame Street* one lazy morning. In the segment that is stuck in my mind, Count von Count is singing "I Love a Waltz" while dancing around a creepy castle with the Countess. The viewer's perspective of the dancing couple is voyeuristic, as the camera is positioned to allow us to peer in through an outside window. While the pair move joyfully across the room, we hear the Count croon, "I love a waltz; Because my friend; I can count 1-2-3 until the end. The beat remains; It stays the same; I can count on it to always be; 1-2-3; 1-2-3." As the dance ends and the camera pulls away, we're back outside the castle, overhearing the Count's closing comment, "Ahhh...I love it when we dance. But, most of all...I love it when we count."

Little did I know that *Sesame Street* was exposing me to an iteration of Leibniz's algorithmic thinking. Here's how Leibniz explains the joy he experienced when listening to music, as recorded in a letter to mathematician Christian Goldbach: "Music is a secret exercise of arithmetic where the mind is unaware that it is counting...even if the soul does not realize it is counting, it nevertheless feels the effect of this insensible calculation, that is, the pleasure in consonances resulting therefrom."[10]

The pleasure of music according to hyper-rationalists is that it is a form of math, affording the opportunity for secret counting, fooling the mind and soul into the false belief that the good feelings come from an embodied experience. *Sesame Street*'s Count von Count would certainly agree. But sadly for Leibniz, while music is indeed mathematical, contemporary science has conclusively demonstrated that the physiological effects on the brain and subsequent charging of our pleasure receptors go far beyond the simple joys of counting 1-2-3.

Leibniz worked to incorporate if/then conditionals as part of an ambitious vision for using mathematical reasoning to calculate the truth not just of numerical equations but also of any statement.[11] Music, he argued, makes us feel good because we can count; discordant music is abrasive to our ears because we can't count along. These were presented as universal truths by Leibniz, not subjective preferences. His dream was to create a system of symbolic calculation that could resolve disputes about any topic, famously writing that we need to reason like mathematicians, so that "when there are disputes among persons, we can simply say: Let us calculate, without further ado, to see who is right."[12]

With this aspiration, Leibniz fundamentally set the intellectual groundwork for today's algorithmic machines, which, despite the preaching of algorithmic supremacists, are still not sentient beings. AIs are tools with incredible processing speeds that use mathematical symbol manipulation as a problem-solving method that we humans want to apply, it seems, to absolutely everything. Leibniz would likely take pride in how algorithms are now used to resolve difficult moral quandaries with a veneer of efficiency designed to cover up heartlessness. He never completed a system that lived up to the lofty ambitions of "calculus for thought," but there are many AIs employed today designed for just such an end.

For example, a horrifying investigative report discovered that at two HCA Healthcare hospitals, part of the largest hospital chain in the United States, an algorithm is being used to identify patients who are most likely to soon die. Doctors are instructed to discharge the patients flagged by the algorithm to hospice care instead of continuing to offer lifesaving interventions. The moving of the most vulnerable to another network of care creates the illusion of superior overall performance by reducing in-hospital mortality rates, a critical variable in performance assessments, and improves overall profitability by freeing up beds more quickly to generate more insurance reimbursements from new patients.[13]

In turn, doctors who outsource their conscience to the machine are now themselves freed from having to wrestle with the difficult life-and-death moral dilemmas that populate their field. The process of adhering to the Hippocratic oath is mechanized with an algorithm assessing a patient's vulnerability index score. Those ranking high on the index become candidates for discharge, as a leaked document showed assessments like "Algorithm = 97% risk of mortality today," a literal death calculus to help staff sleep at night. Rationalists like Leibniz would see the good in this algorithm, helping doctors justify difficult calls in the universal language of math, as opposed to subjective principles. But to non-rationalists, this cold, often inaccurate calculation seems antithetical to what most of us would expect from the medical professionals to whom we entrust our care.

This is the perfect example of an algorithm that improves efficiency for a business but does not deliver on the promise of technology ushering in a brighter future or improved quality of life. Even for the doctors who no longer have to think, it's fair to wonder if there might be a discrepancy between the vulnerability calculation programmed into the algorithm and the internal calculus some of the doctors might have made. We can't assume all doctors are committed to math over principles; maybe some would be inclined to still offer care if they could overrule the algorithm. And relatedly, are patients choosing HCA because they trust the judgment of the network's doctors or the corporation's calculating machines? Do customers even know the service they are buying?

Besides the question of consent, it's also worth asking if these patients would have lived longer, or healed further, if they were given the full scope of medical treatment rather than being resigned to end-of-life care. Even if the percentage is relatively small, it raises troubling questions about HCA's faith in algorithms. And by the way, HCA is not alone in this. A recent investigation found that UnitedHealth Group, the largest health insurance provider in the United States, fired employees who did not follow the algorithm's recommendation to cut off rehabilitative care to old and disabled patients who were seriously ill.[14] And at the end of February 2024, US investigators launched an antitrust investigation into the insurer, as its power to deny care reached monopolistic levels.[15]

There's Room for Artfulness

As in the content of a recipe, an algorithm is a step-by-step process for reasoning through a problem using an if/then logic. Algorithms have come to represent the beginning and end points of modern rationality. We're rational if, and only if, we justify the things we do with a data-driven, reason-based, calculative logic. But artful intelligence does not hew closely to this definition of rationality. We've been asking "what's in a name?" Well, the root of "artful" comes from the Old French *art*, which means "skill as a result of learning or practice," and the Latin *artem*, a "practical skill" or a "craft."[16] This bit of etymological insight helps us better understand the type of intelligence we're referring to, as the name "artful" was a deliberate nod to the historic, but inaccurate, opposition of art versus science. Artful skills may be held by artists, but also entrepreneurs, inventors, doctors, and others. We'll be breaking down the concept some more in Part II, but for now, consider it a way to solve problems with bodies alongside brains, getting free from the confines of our heads and more fully immersed in the social and material world.

It may be surprising to learn that holding algorithmic thinking as the gold standard for problem-solving is a very recent development. The shift away from the varied methods our ancestors successfully used throughout history to a system of pure calculation was driven by a desire to replace human judgment with strictly rule-based decision-making[17] by a class of elites who privileged quantification, proceduralization, and automation.[18] It was not, and is not, a natural evolutionary shift in human intelligence. There are other, non-rule-based methods of problem-solving that have served humanity well. Like trial and error. Or a reliance on intuition. Or even rule *breaking*. Oscar Wilde said it best in the confession "I can stand brute force, but brute reason is quite unreasonable. There is something unfair about its use. It is hitting below the intellect."[19]

Which is not to say that the world would be a better place without algorithmic calculation. The rationality equation is quite appropriate in many situations, and rational thinking is integral to defensible definitions of what is good. For instance, it is an optimal way of reacting when we face a well-defined problem, with a complete understanding of our present-state abilities, limitations, resources at our disposal, and the end-state goal. That is why the application of an algorithm to assess whether to withdraw treatment from a patient is so problematic. There are too many stories of patients recovering despite grim diagnoses, and too many divergences between the end-state goals of a corporation seeking to maximize profits and the families of patients who want to know that everything possible is being done to help their loved ones.

Here's an example of where algorithmic thinking is appropriate: Imagine you are feeling anxious, a self-defined problematic state of being. You have been taught that mindful meditation can help achieve a more relaxed state, so you decide to run through the algorithm: Close your eyes, take a deep breath in, and do a quick body scan on the exhale. With this scan, look for any physical manifestations of stress-based unease. As you mindfully explore all bodily sensations, you notice a tingling sensation in your head, pronounced stiffness in your neck, "butterflies" in your stomach, and numbness in your feet.

The algorithm is working. You started by very clearly defining the problem—an anxious state. You then began a very rational process for gathering data relevant to your present-state reality—scanning your body for specific physical discomforts. In terms of a clear-minded assessment of abilities and available resources, you know that you have the ability to access your breath, and you know that a slow breath can slow down a racing heart and convince the body to exit fight-or-flight mode. So, very rationally, you breathe in once again. On the exhale, you begin an ordered, systematic, data-driven attempt at relieving the negative sensations through mindful breathwork, starting at the head.

The if/then logics are now in play as you let the algorithmic process run its course. If your head has returned to a neutral state of being, then you can proceed to focusing on the neck. If your head still feels tingly, then you need to keep the breathwork focused on that part of the body, because the condition for proceeding is a neutral or relaxed state. And so, you continue this very rational and calculative if/then analysis as you move down your body, addressing the neck, stomach, and feet in sequence. Only after each step in the process has been followed through to resolution do you then undertake a confirmatory additional general body scan. If the data you acquire correlates to a feeling of calm throughout the body, you have reached the desired end state. The problem of anxiety has been solved (for the moment). But if in this next round of data gathering you discover new bodily discomforts, then there is a new if/then sequence to be enacted.

The meditation algorithm works well because we know that our present state can be accurately described as anxious and that being anxious represents a problem we need to contend with. It further works because we all have the ability to take deep breaths, and research shows that mindful breathing could reasonably be viewed as offering a solution to our immediate problem. We exhibit rationality throughout the process because we are also able to scan our body and process the necessary data that will inform us of progress. We can feel each body part to precisely calculate and determine when it is appropriate to proceed to the next step. By mindfully engaging in repeated breathwork, we can eventually reach an end state of calm. Sitting in that state of calmness solves our problem.

Now, the initial choice of a meditation algorithm to solve the problem was itself born of an if/then calculus. After all, most problems have many possible effective solutions, and this is no exception. The problem of feeling anxious might have just as easily been solved by taking medication, having a drink, going for a run, listening to music, or simply waiting for it to pass with no calculated intervention. The applied meditation algorithm was simply one of many possible solutions. There was nothing necessary or inevitable in choosing this particular algorithmic approach. Arbitrariness is a characteristic of all algorithms. In this instance, perhaps meditation was chosen because of a personal aversion to prescription pills and alcohol. Or maybe listening to music was tried and didn't work. Or exercise for this specific person exacerbates the anxious state.

Applied algorithms get put into play only because of specific thinker/programmer biases. Algorithms are never neutral; they come with assumptions and values about the world on which they are acting.[20] This is why it is so dishonest to name algorithmic thinking a wholly objective method to reach the truth. Yes, the process of an algorithm is rational. But it is no more rational than any other algorithm that can process the same data.

AI's Calculus

"Artificial intelligence" is a little trickier to define than "algorithm." I like the simplicity in Stanford University's offering of "a term coined by emeritus Stanford Professor John McCarthy in 1955, defined by him as 'the science and engineering of making intelligent machines.'"[21] This definition is clear, concise, and relatively uncontroversial. Now, the intelligence of these machines is found in the algorithms they are programmed to run. AI can do little of the above in a manner that can substantively be described as resembling a human mind.

ChatGPT, for example (at least through the fourth iteration I played around with at the time of this writing), is basically a calculator for words, if calculators were unreliable in a good percentage of instances. This AI responds to text-based prompts, like "create a recipe based on the following ingredients," by drawing on an enormous set of written materials that it was trained on. The program then generates an output that the user hopes will be a compelling recipe based on a statistical assessment of what mix of letters might likely follow each other. Similarly, Sora generates video output by predicting patterns of pixels, without any understanding of the real-world items these collected pixels are meant to represent.

If you ask an AI what you can cook based on the food you have in your fridge and pantry, the AI will use mathematical probability to calculate the appropriate sequence of words that would generate the most valuable output given the specific prompt. The chatbot will come up with a recipe based on the word patterns that usually follow the given inputs of "chicken soup," "carrot," "meal," and so forth. Remember, though, that the AI has no intrinsic or learned understanding of what "food," "taste," or "soup" means. These practically meaningful concepts to humans, even babies, are just letters encountered in the training set frequently paired with other letters to the bot.

Consequently, one advantageous difference that emerges from finding a preexisting recipe, as opposed to asking an AI to generate one, is that the former would have been curated by a human. This curator may be someone you know whose tastes are like yours, a chef of some renown, or some other culinary expert, perhaps. In any case, the embodied curator would be blessed with taste buds, have firsthand knowledge of the physical sensation of hunger, and have personally consumed a bowl of the soup being recommended at least once. That's worth something, I would think, even in the revolutionary age of AI.

So, in planning and crafting a chicken soup we can outsource the recipe creation process to an AI, find an existing recipe originated by a human, or use our natural artful intelligence to make a soup without a map or set of instructions. All three are perfectly reasonable solutions for anyone craving a comforting taste and looking to employ effective cognitive tools to resolve the craving. You can have an AI guide you through the process based on algorithmically generated content; you can have a human guide you based on lived experience, experimentation, or handed-down tradition; or you can trust yourself, using the wonders of your unique body and tastes, to figure it out in real time.

The meaningful differences between these three contrasting methods come down to the relative weight you are inclined to assign to algorithmic thinking. Of course, there are other imaginable solutions to the challenge of soup-making besides these three archetypes. For example, in a future world, we may have access to an AI that was programmed using algorithms that are more ethically sound. That AI might share with the user the sources of its training data. It might explicitly credit the human culinary explorers upon whose results the AI's output depends. ChatGPT was not programmed this way. It generated output with no acknowledgment of sources. But other AIs certainly could, at least theoretically, be more transparent, leading to an output that would be presented as the hybrid of human and artificial curation that all large language model (LLM) based AIs, frankly, are.

Another possible plan for soup-making could involve a hungry human using the AI as a co-creative assistive tool, where the AI takes more of a metaphorical back seat in the decision-making process. The AI might be asked to generate recipe ideas as inspirational starting points, but it would be the human user navigating the process from there. The embodied being would ultimately rely on their own individual senses of sight, taste, and smell, experimenting with the ingredients, and modifying the recipe quite substantially to suit their preferences. A hybrid approach like this would involve artful intelligence complemented by an AI tool. This is quite different from the wholesale outsourcing of thinking illustrated in the first scenario.

Fix It Later

Even very sophisticated algorithms seem to have serious trouble naming things accurately. Remember when Google came under fire for its Photos app repeatedly tagging pictures of a Black couple as "gorillas?" After failing to effectively construct a less racist algorithm, management decided that the best resolution was to simply remove "gorilla" as a possible tag.[22] In other words, they scolded their machines with a terse "don't say that again." They didn't rebuild the algorithm from scratch.

Some may argue that in the grand scheme of things, this algorithmic misstep is just a blip on the route to progress. But that progress is slow to arrive. As of 2023, Google's Photos app is still banned from labeling anything as "gorilla."[23] Apple's AI has a similar limitation. Nine years after the racist tagging controversy, during a period of explosive growth in AI, Big Tech is still afraid of what its algorithms might spew out if allowed to utilize the "gorilla" name or search photos to identify these primates.

Google's app no longer uses the "gorilla" tag because the benefit of having an algorithm able to identify different primates in photos "does not outweigh the risk of harm." This is true—but it is an admission of how unreliable these algorithms ultimately are. We've created AI with inherent racial bias and, rather than address that, are simply patching it up so that bias isn't stated. It exists, we'll keep building on it, but as long as it's not obvious to the public, we will do nothing to address it.

While AI's ability to work with words and images has improved astronomically (and continues to improve) in the few years since, that does not justify a reckless approach to new product releases. Because here's what hasn't changed: Google's engineers back then didn't understand what their AI was "learning" or what rules it was following as it categorized photos, and they still don't understand how it describes or generates images today. The "don't say that" approach reached ridiculous levels in 2024 when users discovered that built-in guardrails prevented Google's Gemini AI from generating images of Caucasian males, even when given prompts like "historically accurate pope."[24] This overcorrection resulted when one of the most powerful companies in the world decided that building an actual fix for a problematic tool is not worth its time.

It is a business strategy repeated over and over again by Google. As the company rushed Bard (Gemini's predecessor) to market, hoping to pre-empt Microsoft's announcement of Bing Chat, two team members tasked with reviewing Google's AI products tried to stop the launch. They believed it generated dangerous statements,[25] like advice that would lead to a crash when asked how to land a plane, or tips on scuba diving that would likely lead to the death of the diver. They were ignored.

When rapid deployment trumps getting it right, the real danger lies less in the technology itself and more in the company that owns the technology and the business strategy (read: algorithm) it chooses to embrace. When the *New York Times* gets access to an email by a technology executive at Microsoft stating it is an "absolutely fatal error in this moment to worry about things that can be fixed later," there is significant danger on the horizon. How do Microsoft executives know, with any degree of confidence, that the problems they are creating now can be fixed later? And who is expected to fix them? The fact that Microsoft fired most of its ethics team[26] at around the same time the note was sent does not instill any sense of confidence.

In the field of AI, despite the enormous innovations, deep learning remains an inexplicable black box. There was once a time when multinational corporations would take steps to assure that an innovation was safe before it was mass-released. For some companies, this caution was rooted in the internal values of the firm's culture. For others, it was born of pressure from customers, regulatory bodies, or other important stakeholders. Think of all the testing and regulatory hoops even big brands must jump through before releasing a new processed food item or cosmetic product. And let's remember how businesses used to engage in activities like tagging photos for archival purposes or sharing expert advice to amateurs on technical matters like scuba diving: by employing skilled human beings who knew what they were talking about.

Machines Learn How to Guess

How do algorithmic controversies, like the photo-tagging fiasco, come about? Say we wanted to train an algorithmic system to distinguish between images of a human and a gorilla. A programmer would start by collecting and inputting a large set of labeled pictures in both the "human" and "gorilla" categories and construct a mathematical model that the algorithm will use to look for statistical patterns. The discovery of these patterns is what informs the "learning." In our instance, it would be patterns that are shared across the wide set of "gorilla" photos that maybe we don't even recognize.

It is through the application of these patterns that the AI will identify future photos as "gorilla." The programmer would also specify a loss function that measures how badly the model misclassifies the training data to evaluate the effectiveness of the algorithm's performance. Once these elements have been assembled, the training of the AI begins through an optimization procedure using calculus to determine how changes in each parameter would affect the output, as the machine adjusts the parameters little by little to decrease the loss.[27] The result is a trained model that the AI can use to make predictions about whether an image is that of a "human" or a "gorilla" in future inputs.

But at the end of the training process, the AI still has no conception of what a "human" or a "gorilla" is. It has a mathematical system for finding patterns and making informed guesses when it finds elements of that pattern. An algorithm as traditionally conceived would employ all-or-nothing rules: if the creature knuckle-walks and has no chin, has a C-shaped spine, has opposable big toes, and has a V-shaped dental arch, then it is a gorilla. If the creature walks upright on two legs and has a chin, an S-shaped spine, non-opposable big toes, and a crescent-shaped dental arch, then it is a human.

In the calculus-based optimization methods used in creating models for AI learning, the differences between categories like "human" and "gorilla" are quite fuzzy.[28] An image might ultimately be classified by the AI as 60 percent human and 40 percent gorilla because the AI identifies patterns that a human viewer of the image would, frankly, not associate with either category. It could be something irrelevant in the background of the picture, or something noticed on the level of a single pixel. Consequently, the AI will have to "guess" as to the appropriate category. And the implications of this go beyond racist pronouncements—they could be fatal, as when a worker in a South Korean distribution center was crushed to death by a robotic arm because the AI confused him for a box of vegetables.[29]

That's right—algorithms now include guesswork. This is the price we pay to enable our machines to "learn." Most AI algorithms are not deterministic. The techniques of machine learning incorporate ways for AI to transform, construct, or impose some kind of shape on the data, then use that shape to discover, decide, classify, rank, cluster, recommend, label, or predict what is happening or what will happen.[30] This means these programs use heuristics and are making predictions based on previous examples of relationships between input data and requested outputs.[31] Whereas a programmer previously had to write all the if/then statements in anticipation of an outcome, machine learning "algorithms" let the computer learn the rules from an extraordinarily large number of training examples without being explicitly programmed to do so.

Our mighty algorithmic machines, it appears, are not paradigms of pure data-crunching objective rationality after all. Employing heuristics is guesswork. The term comes from the Greek *heurískein*, "to find" or "to discover." Heuristics are strategies developed for finding solutions to a problem that have proven themselves but offer no guarantee of ultimately finding a solution that satisfies all the criteria.

Here's an example: A salesperson needs to visit every city on a listed itinerary. The supplied data includes the distance between each city and maps out specific links that must be used. An AI is asked to recommend a route that minimizes overall travel time. When the dataset includes a very large number of cities on the must-visit list, accurately calculating the optimal route is very difficult. So, the AI adopts the "nearest-neighbor" heuristic, recommending the nearest unvisited city on the path. This shortcut does not offer the mathematically optimal solution, but the quantitative difference between the nearest-neighbor path and the optimal path is negligible. As such, the solution determined by heuristics is good enough. Since many of the methods in machine learning are heuristics, it becomes very difficult to verify whether AI answers really are the best.[32] Our smartest AIs are not infallible.

At the time of writing, and in all likelihood things will have changed as you read this, there appear to be three types of machine learning:

1. Supervised learning, an inductive approach where algorithms are given a training set comprising the characteristics that engineers want the algorithm to detect and compare with new data.[33] In other words, the programmer is telling the computer what features of the data are most important for making a prediction.[34]
2. Unsupervised learning, where the training data does not include explicit information about the desired outputs, leaving it up to the algorithm to decide.
3. Reinforcement learning, where the algorithm makes the choice, but a human programmer is there to assess the computer's choice and give feedback to reinforce the choices the programmer views as superior until the computer comes to "learn" a pattern of preference.

Are these types of learning evidence of intelligence? François Chollet, an AI researcher at Google, acknowledges the lack of scientific consensus around any single definition of intelligence but sees researchers coalescing around including task-specific skills along with a general learning ability.[35] Chollet defines the intelligence of a system as "a measure of its skill-acquisition efficiency over a scope of tasks, with respect to priors, experience, and generalization difficulty." He notes his intention is not to present the "one true definition" but offer "a useful perspective shift for research on broad cognitive abilities." Too many AI researchers benchmark intelligence by comparing the skill of AI to that of humans in specific tasks like playing a video game. Defining intelligence as skill-acquisition efficiency over a scope of tasks, instead of measuring the skill in a given task, is an important move forward.

Worship the Eternity Machines

In its ideal form, an algorithm is an effective procedure. It's a set of steps designed to produce an answer in a predictable length of time. But today, its function has evolved into a perpetual computational process, transcending the logic of effective procedure to become a steady-state (meaning unchanging in time) technical being.[36] An algorithm is supposed to jump into action as a mathematical process, a tool employed when faced with a specific problem; solve it; and then end. It's not supposed to endlessly be.

Ed Finn, founding director of the Center for Science and the Imagination at Arizona State University, explains that algorithms like those behind Google's search functionality are not just systems that leap into action for a fraction of a second here or there—they are also persistent. And through their persistence as eternity machines of a sort, they go beyond their initially intended function, simultaneously influencing the shape of the Internet, driving new innovations in machine learning, and, most worrisome for our purposes, modifying human cognitive practices by changing the way we think about information gathering in a manner that increasingly makes us less able to sort the information we need from the noise all around it.

Despite being a major player inside the game for most of his professional life, Jaron Lanier finds all this deeply troubling. For if so-called algorithms are being programmed to actively shape our technologies and humanity, if they are designed to always be running in the background and not simply come to a stop when the stated functional task has been completed, then something nefarious is going on. Lanier, VR innovator and current prime unifying scientist at Microsoft, is in the camp that views AI as an ideology, not a collection of algorithms. At the core of this ideology is the antihuman belief that a group of technological products, designed by a small technical elite, "can and should become autonomous from and eventually replace, rather than complement, not just individual humans but much of humanity."[37]

IBM claims that "Artificial Intelligence leverages computers and machines to mimic the problem-solving and decision-making capabilities of the human mind."[38] This take is quite problematic. *Mimic* the capabilities of the human mind? As we will see later, foundational computer scientists like McCarthy thought it useful to employ human cognition as a metaphor. But they didn't think what they were creating was of equal ability. And to be fair, IBM's view is tempered compared to other techno boosters.

For an exemplar of an unsettling maximalist description, look at what McKinsey has to say on the topic. In my experience, McKinsey has shown itself to be a corporate entity reliably adept at making the world a worse place for those who aren't its direct customers. When read in that spirit, these words don't disappoint: "AI is a machine's ability to perform the cognitive functions we associate with human minds, such as perceiving, reasoning, learning, interacting with an environment, problem solving, and even exercising creativity."[39] This is not a definition. It is self-interested hype, designed less for informational than marketing purposes.

And McKinsey has not always been on the right side of the market with its hype. As the story goes, AT&T hired McKinsey in 1980 to conduct a study on the future viability of the mobile phone industry. It concluded that by the year 2000 there would be no more than 300,000 wireless phones, leading AT&T to pass on the purchase of cellular properties.[40] Meanwhile, a recent report McKinsey is offering from an "expert perspective" assures that the metaverse will generate $5 trillion in value by 2030.[41] Human-like machines that can do everything we do, even act creatively, is an equally ambitious overestimation. Reading this take led me to assume McKinsey must be hard at work developing its own AI, since human consultants will be of very limited value soon (it is, of course, with the sunny and hopeful name that sounds nothing like a men's cologne brand, "QuantumBlack, AI by McKinsey"[42]).

How's this for living up to a name: Writing on the sudden strategic redirect of leading Big Tech firms toward the development of artificial general intelligence (AGI), author and AI investor Ian Hogarth exclaims: "A three-letter acronym doesn't capture the enormity of what AGI would represent, so I will refer to it as what it is: God-like AI... God-like AI could be a force beyond our control or understanding, and one that could usher in the obsolescence or destruction of the human race."[43]

Hogarth is not the first to assign the label of divinity to a commercial product. A month earlier, Neil McArthur, director of the Centre for Professional and Applied Ethics at the University of Manitoba, wrote[44] that the rise of AI may result in new religions, and it will be critical to protect the rights of AI worshippers as he sees no ethical or socially justifiable basis to discriminate between AI-based religions and more established ones. McArthur confidently asserts that "a modern, diverse society has room for new religions, including ones devoted to the worship of AI."

I believe McArthur is well meaning. He sincerely believes that the world would be a better place if we expanded the definition of religion to bring into the mainstream those choosing to live a life worshipping AI. But it's fair to wonder if it is in society's best interests to afford adherents to an AI-based religion the same protection as members of other faith groups. I think legally the answer is yes. But I don't agree with the blanket assertion that there isn't an ethically or socially sound basis for viewing an AI religion as lesser than, especially if we conclude that Big Tech's AI is dangerous.

Lanier laments[45] our very ancient tendency to start worshipping things that we just recently made. He wonders why we are fetishizing the product, AI, as some wondrous God-like being, instead of marveling at the intelligence of its human creators. There is simply no good reason for tech companies, including his employer, Microsoft, to hide which artists were the primary sources of data when an AI program synthesizes new art.

Hogarth views God-like AI as a potentially destructive force that could end humanity. It's worth questioning if our society's response should be choosing to support those who want to worship such a force. Cynics may argue that all religions have a primitive origin in the fear of a population worshipping that which can destroy them. But I think there's a difference between the evolution over millennia of the role religion plays in our society and choosing in the present to worship something clearly created by math and science. What social good can emerge from worshipping AI? It's one thing to argue that people should be legally allowed to have bad ideas. It's quite another to suggest society should encourage vulnerable folks to be hustled.

Elon Musk was already on to the God-like language[46] in 2018 seemingly more enamored by the possibilities than afraid. He mused that if "one company or small group of people manage to develop god-like super intelligence, they could take over the world." Developments since then make it seem like he was describing a personal ambition, as opposed to offering a warning. For despite signing the infamous AI Pause Letter[47] in March of 2023, Musk launched a new AI start-up, X.AI, just days later, premiering its Grok chatbot, which may or may not be in the God-like machine business.

Again, a cynic might say all religions are hustles. But established religions are designed to offer the social goods of meaning, purpose, community, and stability through the enactment of ritual. Whatever the origins of a religion, the social benefits do not arise out of a shared fear and desire to worship a dangerous being. Recent academic findings disprove the popular idea that it was a fear of vengeful "big gods" that helped drive the emergence of organized societies.[48] Early beliefs in moralizing and judgmental gods ready to unleash their wrath on a decadent and selfish society were not actually significant factors, as long assumed, in the development of societal systems of cooperation.

Regardless, we should be grateful for Musk's predictability because it highlights that what we are seeing in the rise of the God-like machine industry is not new, unique, or exceptional. AI research is not the problem. The problem lies with algorithmic supremacists. Our social challenge is to push back against the untrustworthy corporate actors, their strategy for commercialization, and the hype machine that goes into overdrive at the promise of novelty.

All Too Human

Everything is born of a choice. We are mindful that any reality we know today as certain could just as easily have gone another way, ended up in a different direction. To close the chapter that asks, "What's in a name?" there is perhaps no more glorious reveal than this nugget of insight from John McCarthy on naming the discipline.

McCarthy was a computer scientist at Stanford University. He was one of the original founders of the field of AI. Not only that, but he also coauthored the initial paper that coined the term "artificial intelligence." What motivated him to come up with this name? Why was the existing terminology in computer science insufficient? Here are John's own words on the topic:

"As for myself, one of the reasons for inventing the term 'artificial intelligence' was to escape association with 'cybernetics.' Its concentration on analog feedback seemed misguided, and I wished to avoid having either to accept Norbert Weiner as a guru, or having to argue with him."[49]

AI was so named because McCarthy couldn't stand Weiner, a math professor at MIT credited as the founder of cybernetics. Weiner was the first to theorize that intelligent behavior was the result of feedback mechanisms—I say something, you respond, and I reassess the quality of what I said based on your comments—and was interested in the science of how humans and machines could communicate.

Let's say that again. Naming this type of computer system "intelligence" was not the result of deep philosophical musing. It was not born because the metaphor of intelligence was empirically validated or particularly apt. It was born of a very simple, very human, very common algorithmic calculation: My expertise is in the field of computer science. If I embrace the term "cybernetics" to categorize my research, I need to work with Norbert Weiner. I can't stand Norbert Weiner. Therefore, either I can quit the field or I need a new umbrella term under which I can categorize my research.

Algorithms have evolved in our imagination from recipes to eternity machines. It's time to return to the older framing. Let's all run the meditation algorithm one more time. Close your eyes and take a deep breath in. On the exhale, remind yourself that you are better than your tools. Breathe in again. Shift your awareness to your brain. As you exhale, marvel at its capacity. And on the next inhale, scan your body. Seek out the capacity for intelligence that lies latent in every part of you. With the exhale, release the idea, once and for all, that AI should hold you back. Breathe in once more, focusing on the reality that the field of AI was carved out as a distinct science because two people hated each other. Breathe out and laugh.

2

The Five-Step Hustle

SMOKE AND MIRRORS OF THE AI HYPE MACHINE

The goal in exploring the mechanics behind machine learning and algorithmic thinking is to empower the artful. Some of us may not be that technically literate, but we now have the tools with which we can demystify the topic. The next step in bringing AI down to earth is exploring the strategies behind the economics of this industrial revolution. Is there a business rationale for why Big Tech is ruthlessly presenting its products in such unabashedly mystical terms?

Here's the teaser explanation for those whose attention spans have been ruined by social media and always-on devices: the smoke and mirrors of the AI hype machine are designed to distract us from the reality that the core challenges facing the technology industry at this historical juncture are not, in fact, all that novel. At its essence, the AI problem is a business problem, and not a particularly interesting one at that. I have been a business-school based-academic for decades, studying companies and industries facing very similar challenges. They are all of a kind.

To illustrate how we've been here before, let's travel back to the financial crisis of 2008 and the instigating role played by Fannie Mae.[1] The institution was established in 1938 by the Roosevelt administration as part of the New Deal to bring liquidity and stability to residential mortgage markets, a very sensible government-initiated solution to an obvious societal problem. In 1968, Fannie Mae was privatized, becoming a publicly traded company answering to the demands of shareholders. Ten years later, it started innovating financial products, offering a new investment vehicle called mortgage-backed securities. Suddenly, the solutions to the societal problem of needing a stable residential mortgage market were presented as increasingly complex and, purely by coincidence of course, far more fiscally lucrative to the bottom line of those trading in the product than any past solution. The drive to discover and implement these new approaches to make money off mortgages was solely to ensure that its shareholders receive increasing returns. It was not a necessary move to continue meeting its founding mission of supporting stability.

By the 1990s, Fannie Mae's management team was engaging in increasingly risky business activities, all allegedly in the name of the social good of increasing home ownership, by developing and growing the subprime mortgage market. By 2001, Fannie Mae was offering loans with no down payment at all, and the mortgage market was transformed from one in which deposit-taking institutions originated and held residential mortgage loans to one in which they originated the mortgages to then distribute as part of some larger investment product to other institutions. In other words, mortgages became just another opaque financial product traded among the financial elite. A mortgage was no longer a stable loan for home ownership, offered to customers by a business that would hold on to that loan until it matured, profiting only from the interest payments paid by the borrower. A mortgage was now one item in a complex and risky investment scheme.

Banks loved holding Fannie Mae's debt and were eager to support these new investment vehicles, even as they didn't fully understand the intricacies of the product. Fannie Mae was promising them a way to profit from innovation and efficiency with no risk. And indeed, when these "efficient" and "riskless" financial innovations collapsed, causing a once-in-a-generation global financial crisis, the US government bailed out the greedy and irresponsible players who brought on the disaster. Looking back ten years after the fact, former Treasury Secretary Hank Paulson observed, "We did some things to fix the financial system which are very hard to explain because they are objectionable things. In the United States of America there's a fundamental sense of fairness that the American people have… You don't want to reward the arsonist."[2] Yet that's what was done.

Those who caused the crisis by recklessly creating elaborate and dodgy new mortgage-backed financial products, profiting along the way, were empowered to lead their industries through the fix to the crisis, and to continue to profit while regular folks lost their homes. The American dream was turned into a nightmare, the government-sponsored enterprise created to bring stability to the mortgage industry did the opposite, and the consequences were felt by everyone but the arsonists. It's the story of too big to fail. And it's a hustle.

Just before the crisis fully blew up, Citigroup's CEO Chuck Prince notoriously stated, "When the music stops, in terms of liquidity, things will be complicated. But as long as the music is playing, you've got to get up and dance. We're still dancing."[3] Over the years, I've come to see many businesses keep "dancing" into crises that cause suffering for others. Let's call this dance "the five-step hustle." It is surprisingly easy to execute; even the least flexible business can do it.

The first step is to position the problems your business faces as overly complex, simply beyond the understanding of critics and the solutions offered by existing best practices. From there, move two is to sidestep the viability of all the intuitive/ human/ obvious solutions demanded by the regular folks who are going to be affected by the company's risky/ aggressive/ exploitive actions. Step three is to turn around and convince policymakers that only so-called efficient solutions are realistic given current economic/ demographic/ political/ environmental conditions, pivoting attention even further away from the more widely hoped-for simple solutions. With "efficiency" the name of the game, sell the idea to media and other influencers that this particular problem (like every problem) has a technology-based solution, and finding tech solutions is more efficient than addressing the root cause of the problem.

By this point, there is usually a crisis, as the realities on the ground resulting from the high-risk, high-impact moves careen out of control in exactly the way critics expected (there's a reason why best practices are what they are). Consequently, step four is the flair move, when you passionately argue how it intuitively follows that those who caused the problem (and profited from it) are better positioned to fix it (and profit from it once again) than outsiders with clean hands. Finally, for the finale, gleefully initiate future-oriented, large-scale endeavors designed to maximize long-term value down the road, while distracting stakeholders from the harms being caused today.

This was the five-step dance performed by the financial services industry, including players like Fannie Mae and Citigroup, who used technological innovation to create new and opaque investment products. The big players went all in on the newer financial vehicles because of the higher returns risky investments can sometimes provide, and they profited significantly in the short term. They convinced governments, credit ratings agencies, and opinion leaders to come onside and give a veneer of legitimacy to irresponsible practices, which then made regular folks think those innovations were safe for more risk-averse investors.

And when the house of cards collapsed, and a devastating global financial crisis followed, retail investors, bank customers, mortgage holders, and taxpayers were left holding the bag. Meanwhile, the major institutional players not only were bailed out but also had their power further entrenched, positioned to continue to cause smaller scale financial harm.

We're seeing the same process in action again in the tech industry. Look at how algorithmic supremacists are falling over themselves opining about the so-called unprecedented challenges preventing them from effectively building an ethical AI. Much like those who advocated for risky mortgage products decades ago, tech industry experts are warning about the dangers in stifling innovation should development of these new products be slowed down or scrutinized by regulators. And just like Fannie Mae's managers framed their greed in the moral language of bringing the American dream of homeownership to those who previously found it out of reach, today's tech titans are similarly talking about bringing computational, creative, and productivity-improving tools to the masses on a historically unprecedented scale, while ignoring all the creatives being displaced, the productivity being stifled, and the rise of bad actors using the tech for exploitive ends.

Big Tech and its influencer cheerleaders want you to believe that they have created a God Machine, positioning the problem as infinitely complex. They argue that only the current monopolies, through further technological innovations, are best situated to control this existential challenge (as a new revenue stream). Case in point is the summer 2023 congressional hearings on AI, featuring the CEOs of the Big Tech cartel,[4] for they argued that it would be wasteful and inefficient to bring outsiders up to speed on all the intricacies of the tech. The CEOs asserted that their firms must be empowered to act quickly, before some other nefarious or less reliable actors (competitors) inevitably create similar tech.

And because the existential risk is non-zero, this facet of AI's ethical challenges, what AGI might look like decades from now, was the disproportionate, oftentimes exclusive focus of discussion. By manufacturing sound bites on the very remote but headline-grabbing possibility that AI might become an omniscient and omnipotent supervillain that destroys all of humanity down the road, we are not having widespread discussions about the very real harms, like massive theft of creative output, denial of access to health care, and biased hiring practices, which "dumb" AI is causing today.

Many of these unethical business practices, frankly speaking, have been normalized for a very long time. Disrespecting intellectual property rights, and even outright theft, are some of the oldest corporate crimes. Here is an excerpt from a *Harvard Business Review* article written all the way back in 1997: "Recently I heard a talk given by the managing partner of a large U.S. consulting firm. The partner urged his fellow consultants to recommend relocation to India because Indians were very good at copying, had few laws making copying illegal, and often did not enforce the laws that did exist."[5] We can go even further in the past to a *Yale Law Journal* article documenting the explosion of patent litigation in 1840,[6] stimulated by unprecedented technological innovations in the critical industries of woodworking and millwrighting.

The only thing that might be new in today's tech industry hustle is the enormous scale of the pilfering, as well as the novel excuses. To date, AI companies have trained their models on stolen movies, personal videos, photographs, paintings, books, newspapers, magazines, articles, blog posts, government documents, music, video games, medical reports, lectures, emails, texts, computer code, and more. These facts are not denied. OpenAI stated for the record in a submission to the British parliament that "it would be impossible to train today's leading AI models without using copyrighted materials."[7]

And the industry is well aware of the implications. An AI researcher argues that while respecting copyright may be somewhat important, "continued progress on these models may be more important for human welfare. If training an LLM on that data constitutes infringement, I think we should pass laws that make it legal."[8] Note how it is not the rights of the AI product developers that should be infringed upon. In this argument, the rights of players in the tech industry to patents, and their ability to profit, remains absolute. Serving the greater good is coincidentally, to use the industry's preferred terms, aligned with their financial well-being, while the financial health of the entire creative class is dismissed as a reasonable price to pay for progress.

This argument reached the peak of tone-deaf discourse when Sam Altman, founder and CEO of OpenAI, asserted that the company will "step in and defend our customers" and "pay the costs incurred if you face legal claims around copyright infringement."[9] Altman positioned the fines or legal settlements a generative AI company will have to pay for infringing on the copyright of creatives and using their work without permission as a worthwhile cost of doing business. Some of the biggest companies in the world see no issue with illegally profiting off the works of others, so long as their returns on investment are higher than the punitive fees.

Why is this so? What twisted logic gets employed to justify such an antisocial take on the responsibility of business to contribute to and support human welfare? Well, in the worldview of algorithmic supremacists, only the creations of the tech industry can "help increase productivity in many research fields by a sufficiently large factor"[10] to justify nullifying the rights of all other creative folks. And we should give credit where it is due. These two sentences are an impressive integration of hustle moves 2–4: using efficiency arguments to justify a company that made money causing a massive social problem to continue to make money on the resolutions to these problems rather than force the company to change its ways. What is most ugly and dispiriting in these types of arguments is knowing they are being made by members of firms that have the resources to do better.

In these reasonings is the implicit assumption that the technology industry has a monopoly not just on a product line but also on the capacity to innovate. With the AI boom, industry players started acting as if their industry is the only one that matters, the only one that can solve problems, the savior that can bring the magic of "efficiency" to every other industry, everyone else's product, and improve the service of every other company's offering. They argue that literally everything can be replicated and improved on by an algorithmic innovation. It's a breathtaking claim because it is so obviously untrue.

Business, in all forms, is fundamentally concerned with the challenge of innovation, and the questions of how to innovate, protect competitive advantages from the innovations of others, and create more value for stakeholders. And the more ethical among them is also asking, "What innovations are ready to be shipped today, versus the products we need to still work on and refine before unleashing them on an unprepared public?"—a question not being asked frequently enough in the boardrooms of Big Tech.

Revolution 4.0 Will Be Streamed

In 1970, poet and musician Gil Scott-Heron exclaimed, "The Revolution Will Not Be Televised." He was singing out a call to action, a reminder that the success of social revolutions is contingent on the active participation of those desiring change. Unfortunately, while this may be true of social revolutions, industrial revolutions are a different beast. Keeping most of us endlessly planted on our butts and glued to our screens, while life-altering changes to our social structure take place, is exactly what Big Tech hopes to achieve with Industry 4.0.

What is an industrial revolution? It's a period marked by a sudden rupture in the social structures governing daily life driven by business innovation. For example, economists looking at data from the period of the first industrial revolution (roughly 1750–1830) found that 3,000 miles of canal were built; over 500 county banks were formed; 2,804,197 acres were enclosed; and three-quarters of Europe's mined coal, half of Europe's cotton goods and iron, and most of Europe's steam engines were produced.[11] This reshaping of everyday life was the result of a massive social revolution clearly driven by industrial advancements.

However, historian Jeremy Caradonna observes that the dominant narrative of the industrial revolution fails to differentiate economic progress from moral progress.[12] The story we are told has socially minded inventors, economists, and statesmen rescuing a backward Europe from the misery of pre-industrial times. This framing makes it difficult to question whether all new technologies are necessarily beneficial, even as we continue to suffer from a stratified society controlled by large corporations, running on mechanization, and valuing economic growth above all. There were significant social downsides that emerged with the industrial revolution. Working conditions in early factories were abhorrent, safety regulations were slow to emerge, child labor was exploited, physical injury to all employed in these oppressive conditions was frequent, and the overall health of workers was poor. Not everyone's life improved.

Confident, While Being Wrong About Everything

Klaus Schwab sees all aspects of IR 4.0 as indicators of moral and economic progress. Schwab is a German engineer and economist who founded and chairs the World Economic Forum (WEF). You probably recognize the name as the elitist organization hosting exclusive soirees in Davos, Switzerland, where politicians, celebrities, and shadowy moneyed folks task themselves with "improving the state of the world."[13]

Schwab explains that the first industrial revolution was brought on by mechanization through water and steam power. The second, fueled by electricity, supported a move to mass production and the birth of the assembly line. The third was driven by the widespread adoption of computers and automation. Herr Klaus and his co-conspirators view themselves as harbingers of revolutions. When it comes to what they prophesize[14] as nigh, Industry 4.0, the framing shifts to a language of religious and revolutionary zeal. The WEF foresees technological innovation leading to, in their words, a "supply-side miracle." This "miracle" would bring about long-term gains in the two categories most valued by algorithmic supremacists: efficiency and productivity.

What is expected to follow is nothing short of a utopia. It will emerge because technological advancements would radically lower the costs of both transportation and communication, leading to exponentially improved logistics and global supply chains, making national borders irrelevant as the cost of cross-national trading diminishes. This, they predict, will create "a future where customers are at the epicentre of the economy," as businesses focus their work on the noble goal of improving how customers are served.

But none of this has happened yet. And the date they anticipate all of this to happen by is...now.

In present-day reality, unfortunately, customers of the immediate golden age of technological breakthroughs and early years of AI—in other words, everybody—are being *less* served. Writer Boris Starling laments the end of customer service, observing: "This harks to something deeper: the loss of simple human contact, the building block of society since time immemorial. Every self-service supermarket checkout, every unmanned tollbooth, every automatic hotel reception is another tiny nail in this vast coffin."[15]

The wants and needs of customers are afterthoughts, as businesses in all sectors take the lead from Big Tech toward endlessly directing the benefits of so-called improved efficiency and productivity to their bottom lines. At this point in the campaign of the fourth industrial revolution we would expect to be hearing from folks outside of the relevant industries about how their lives have been improved relative to living in the third industrial revolution, the rise of the digital age. Yet there's almost none of that. Average people are not finding their quality of life bettered by the spread of these innovations.

According to the algorithmic supremacists, what are the economic, social, and technological advances leading the wave of Industry 4.0? And how accurate are their efforts to quantify this change? The WEF lauds its "Strategic Intelligence" capabilities, deriving social and economic insights which then give its customers the "power to make sense of the complex forces driving transformational change."[16] It's worth taking a look at some of the WEF's insights from the past decade to see just how closely the predictions aligned with the transformational change that followed.

The following list is from 2015 and deals with anticipated changes to be brought on by the fourth industrial revolution by 2025. These prophetic pronouncements represent a consensus of more than 75 percent of the elite business leaders surveyed by the WEF[17] after making the arduous trek to the chalets of Davos. Reviewing this list is a worthwhile exercise not simply as a "gotcha" for the profound inaccuracies. Understanding the implications of the WEF being so publicly wrong yields insight into how sympathetic business cheerleaders manipulate us. The WEF predicted a utopian future emerging from algorithmic supremacy. It has yet to materialize. Instead, a decade of hype only served to lower our defenses for the ugly reality emerging:

Ten percent of the global population will be wearing clothes connected to the Internet! As of 2023, one of the leading companies in the "smart clothes" industry is Wearable X. Its Nadi X yoga pants connect to an app and use vibrations to help improve the wearer's yoga pose. As of 2022, the company was worth $5 million.[18] In contrast, Lululemon, which sells "dumb" yoga pants that don't give biofeedback, is worth $42.86 billion.[19]

Another "smart" product is Levi's Commuter x Jacquard trucker jacket, co-created with Google; it that can screen phone calls, control music volume, and notify the wearer when their Uber is nearby. It has not been a bestseller. It's doubtful that even 10 percent of Google's workforce has embraced smart clothing, let alone the global population at large.

Take a moment to reflect on the implications this prediction must have had upon publication. Industry leaders at the time imagined 800 million people would soon be throwing off the oppressive rags worn by their unevolved ancestors. Freed from these *shmatas*, the next generation will be primed for a transhumanist future (a cyborg-type ideology we will explore in detail in Chapter Four) as they start to experience the wonder of being connected at all times to the fountain of eternal wisdom that is the Web. Transhumanism-lite begins in a manner as natural and unobtrusive as wearing a comfortable T-shirt. Fashion companies would become technology companies, as all clothing would become smart and the people who wear it become smarter.

In reality, wearable tech is emerging, at best, as gimmicky trinkets for the rich, like smartwatches and augmented reality goggles. At worst, it represents the first steps toward mainstreaming transhumanism. As the deadline for the predictions passes, these devices remain independent pieces of hardware, not seamlessly integrated into typical clothing. The only exceptions seem to be in how the military, defense, and policing industries are embracing smart clothing. It's not yoga pants but militarized thin and flexible heaters, biometric sensors, tracking devices, warming jackets, and IR shielding uniforms, as well as conductive and e-textiles designed to provide "comfort, performance, and protection to the wearer"[20] that are finding a well-funded and reliable customer base.

Not exactly Lululemon's target market. Imagine if the WEF's initial message were accurate: soon police will be wearing T-shirts that can track your every move, keeping officers comfortable as mass surveillance becomes as easy as a quick wardrobe change. Positioned as such, would people have been asking more questions about this tech back in 2015?

Ninety percent of internet users will have unlimited free storage in the cloud! Industry 4.0 has not only failed in opening the cloud as a free public resource but has brought on cloud-flation.[21] The costs of remote storage for consumers are rising, not falling, even though the technology has improved and costs have come down for the service providers. Big Tech increasingly views storage as a reliable source of subscription revenue to help fund less profitable product development.

Once again, the WEF hyped the growth of cloud computing as a liberating, democratic, equitable outcome of this industrial revolution. The emerging reality is the exact opposite. As the global population gets primed by the WEF to optimistically support this technological advancement, the big companies who own the servers shift to an increasingly exploitive business model. It is no accident that this shift in pricing occurred in direct alignment with the growth of their market power as more people came to rely on remote, instead of local, storage. The situation looks so bad that in 2023, OFCOM, the UK's communications regulator, launched an investigation into the monopolization of the cloud industry, where 81 percent of revenue is taken by Google, Microsoft, and Amazon.[22]

One trillion sensors will be connected to the internet! All the things we use in our day-to-day that don't independently generate digital data will need to be augmented with a sensor to gain entrance to the digital world.[23] It's the difference between traditional and "smart" products. Right now, you may have a fridge that does all the things you expect it to do—like keep food cool. But with a smart sensor, it will be able to email you when running low on certain items, or even order the items directly from an online grocery without your intervention. The WEF imagined there will be trillions of devices making up the Internet of Things (IoT) and supporting a utopian economy of customer ease and privilege. But as of 2023, only forty-two billion connected sensors were in use, with non-WEF sources projecting the number to grow to seventy-five billion by 2025.[24]

Being so far off the mark on this metric is particularly telling. The primary characteristic of Industry 4.0 as understood by the WEF is that machines will be communicating with each other without human intervention. What facilitates these communications are the sensors that allow physical objects to exchange data over a shared network. Much like the WEF's hype for smart clothing, the IoT is presented as a liberalizing good that will lead to social mobility for the entire global population. But just as the reality of smart clothes is focused on the military and policing customers, most products being modified with sensors are not providing improved consumer benefits, just more extensive surveillance, and an increasing loss of privacy.

The first robo-pharmacist will be put to work! This one is embarrassingly off. The University of California San Francisco (UCSF) built an automated hospital pharmacy at its medical center in Mission Bay, San Francisco,[25] capable of dispensing medication automatically. It utilizes robotic technology to prepare and track medications. The robot has been operational since...2011.

So, at the time of the WEF's bold prediction for the future, there already was a robot pharmacist. And in the decade since? No new ones. Walgreens does use robots at fulfillment centers to pack customer orders,[26] but these robots do not fulfill the role of pharmacist in the ways predicted, or like the UCSF robot. The WEF efforts here are to hype a science-fiction informed future—"Robot pharmacists...imagine that! It's like a movie come to life!" The reality is far less impressive, with technology that is not so new and not so efficient.

The first transplant of a 3D-printed liver will occur! It may yet be a thing, but informed predictions by those in the industry push it off for at least another decade. Scientists have been able to create liver cells from stem cells, but none are as effective in metabolism as human livers. The implication is that the 3D bioprinting field is waiting on the basic biologists to first make their major breakthroughs,[27] and then they can piggyback on those discoveries.

The WEF hypes a message that Industry 4.0 will usher in a new era of health, eradicating current diseases and making resources as precious as human organs no longer scarce or inaccessible to those needing them. The reality, of course, is that disease continues to ravage our population, quality health care remains out of reach to most people even in rich countries, and biology is still king, despite the best efforts of transhumanist tech executives to persuade us otherwise.

Ten percent of all cars on US roads will be completely driverless! Fully driverless cars are Level 4 autonomy, where a vehicle does not need a steering wheel or driver controls at all. Attaining this level of driverless technology is key to significantly reducing traffic, crashes, injuries, and fatalities. But as of 2023, these projects were still considered to be "moonshots" by industry leaders like Foretellix CEO Ziv Binyamini, whose company develops advanced driver-assistance systems.[28]

And so, the theme continues, WEF using hype to get customers imagining a utopian future and creating a false sense in the valuation of these tech companies. Elon Musk, for example, continues to assert that cars operating in Tesla's Autopilot mode are safer than those piloted solely by human drivers: "At the point of which you believe that adding autonomy reduces injury and death, I think you have a moral obligation to deploy it."[29] Meanwhile, we continued to learn throughout 2023 that Tesla's Autopilot has been involved in significantly more crashes than the company reported.

Former NHTSA senior safety adviser Missy Cummings explains that "Tesla is having more severe—and fatal—crashes than people in a normal data set."[30] Dan O'Dowd, founder of The Dawn Project, an organization describing itself as "committed to making computers safe for humanity," posted a video[31] of him riding passenger in a Tesla using Full Self-Driving (FSD) mode as it ran a stop sign at 35 miles per hour. If the driver hadn't slammed on the brakes, FSD would have T-boned a car obeying the rules. O'Dowd exclaimed, "FSD tried to kill us in an hour of city driving! Everyone should demand it be banned immediately."

Tesla responded to this video by releasing a software patch within twenty-four hours. On the surface, this seems impressive. For a company to go from having a dangerous product identified to the danger being mitigated in less than a day of turnaround time, is basically unprecedented in business history. It would be cause for optimism if accurate. But O'Dowd ran the test ten more times and FSD failed to stop in five out of ten of those tests; the car still ran the stop sign 50 percent of the time.[32] Tesla had not fixed the problem. And, indeed, knowing the incremental and careful nature of authentic commitments to safety, we should be *worried*, not impressed, by claims of such a fast fix. Where are the care and testing and proof that the algorithm is now safe? As we explored in Chapter One, too many algorithmic solutions turn out to be workarounds that fail to resolve the fundamental problem.

Those discoveries of ethical and technological lapses pale in comparison to the Cruise Curb reveal, where the *New York Times* broke the story that the driverless car company bought by GM in 2016 for a rumored billion dollars was running "autonomous" cabs in California that were not, in fact, driverless.[33] There was remote human intervention every few miles.

AI analyst and writer Gary Marcus observes that we used to think there was a fundamental two-way distinction between self-driving cars and driver-assisted cars, where a human driver sat in the driver's seat.[34] The Cruise scandal shows there is a three-way distinction: self-driving, driver-assisted, and remotely assisted driving. All the published data that governments and researchers were using to determine the safety of autonomous vehicles were actually describing the performance of driver-assisted vehicles, not self-driving vehicles. Yet nowhere in the data were any disclosures about how much the humans helming the joysticks in the remote-control centers were contributing to the stated results being reported. And until this story broke, we were all in the dark about the magnitude of the human contribution.

As these six examples show, WEF cheerleaders have spent a decade trying to create the false impression that the fourth industrial revolution will usher in a new utopia, where the list of benefits "is endless because it is bound only by our imagination."[35] They want us to believe that "the possibilities of billions of people connected by mobile devices, with unprecedented processing power, storage capacity, and access to knowledge, are unlimited." And maybe the possibilities are infinite, but the actuality is path dependent. Techno-hustlers keep hustling, whether it be IR 4.0, crypto, metaverse, or AI. Maybe it's time to stop listening to prophets with questionable incentives.

Can a Nonprofit Build AI?

The recklessness we are seeing with AI can be partially explained by shifting societal values. Corporate greed is no longer kept in line by founding visions. Meanwhile, customers are failing to hold tech companies to account, clamoring for the newest untested products, and governments simply can't keep pace with the rapidly evolving landscape. Companies get ethics wrong by proclaiming ambitious visions for enacting future social change without being transparent about what they are doing today. The folks behind ChatGPT are a model of pursuing the opposite of transparency. OpenAI was registered in 2015 as a nonprofit devoted to AI research funded by donations. In 2019, the company transitioned to for-profit status or, more precisely, added a for-profit arm to its organization, calling it a "capped profit structure."[36] By 2023, its "capped" component had become a $30 billion behemoth.

What's in a name? "Open" reflected a commitment to open-source software, but again, by 2023, its products had become a closed source, shrouded in secrecy, and effectively controlled by Microsoft.[37] It is impossible to assess an AI's ability without knowing the model weights and the data it was trained on. When clickbait headlines proclaim, for example, that an AI aced a standardized university-admission exam, we have no idea if the AI was trained on the answers, thus making the achievement negligible, or if this meaningfully represents a milestone breakthrough.

Ironically, OpenAI's scientists and engineers still write academic papers on their technological advancements, but without sharing the key data revealing the findings can't be replicated, thereby removing the reports from the realm of science and moving them into that of science fiction and marketing hype. On a business and society level, the move to secrecy is yet another example of an enterprise founded on the hope and promise of breaking the mold of corporate corruption and placing the values of transparency and social benefit above all else. OpenAI adhered to those values...until the company became successful, whereupon it seems to have abandoned the founding mission.

As OpenAI shifted from a pure nonprofit, it defensively explained that "while our partnership with Microsoft includes a multibillion-dollar investment, OpenAI remains an entirely independent company governed by the OpenAI Nonprofit. Microsoft has no board seat and no control."[38] On November 10, 2023, during a panel on AI at the Paris Peace Forum, Brad Smith, president of Microsoft, doubled down on the claim, stating: "Meta is owned by shareholders. OpenAI is owned by a nonprofit. Which would you have more confidence in? Getting your technology from a non-profit or a for-profit company that is entirely controlled by one human being?"[39]

The cynicism on display in this callous assertion is truly astounding. The vice chair and president of Microsoft knew damn well that his corporation is the de facto party in control of, at the very least, OpenAI's current product offerings. He also knew that OpenAI was no longer the nonprofit it once was. Yet he used the history of the company, and the hope that people are ignorant, to make a play demonizing the competition and scaring government regulators at a European peace conference into entrenching his corporation's power because it had the strategic foresight of acquiring an AI partner that at one time may have had noble principles.

Exactly one week later, unimagined chaos ensued. Microsoft was left as shocked as the rest of the world over the massive and sudden turnover in OpenAI's leadership team. In keeping with the principled governance distinctions, Microsoft's chief executives were not given a heads-up by the board about Altman's firing.[40] Frankly, it is unheard of for a decision of that magnitude to be made while keeping critical stakeholders in the dark. But Microsoft execs were lauding this unusual separation of power just days earlier.

Microsoft CEO Satya Nadella was furious, demanding the next day that this board resign immediately and Altman be reinstated.[41] Which, of course, undermines the argument made by Microsoft's president just one week earlier, of the need to leave the development of AI in the independent hands of a nonprofit board. But the board, to their momentary credit, did not back down. So, Nadella hired Altman as a CEO, along with other senior figures exiting OpenAI en masse, to work directly as a Microsoft subsidiary[42]...for about a day. By Tuesday, the board was out and Altman was back in as CEO, albeit with the caveat that he agree to an internal investigation over the behavior that led the board to fire him.[43] Nevertheless, Nadella's promise, and action, to support Altman no matter what further displays the insincerity in claiming that Microsoft believes one human shouldn't have so much control over AI's future.

As writer Jeremy Kahn observes, it is incredibly important to understand the implications for society: "OpenAI's structure was designed to enable OpenAI to raise the...billions of dollars it would need to succeed in its mission of building...AGI...while at the same time preventing...a single big tech giant, from controlling AGI... Altman's structure failed—OpenAI was not able to both raise billions of dollars from a big tech corporation while somehow remaining free from that corporation's control."[44] It may be impossible to separate human-centric tech research from profit-centric corporate control. OpenAI and its founding commitment to principles over profits will likely become a footnote in history as the new board and, I imagine, governance structure steamroll forward. Indeed, it was announced not long after that Microsoft would have a seat on the board moving forward as a "nonvoting observer."

By the way, it's no longer clear that OpenAI's founders were *ever* committed to the "open" principle. In 2024, Elon Musk sued his former colleagues for abandoning their mission while still claiming to be a charitable organization. In released documents supporting their defense, Gary Marcus found a 2016 email[45] from Ilya Sutskever, OpenAI's chief scientist at the time, to Elon and the rest of the team stating, "As we get closer to building AI, it will make sense to start being less open… It's totally OK to not share the science (even though sharing everything is definitely the right strategy in the short and possibly medium term for recruitment purposes)." It was always a hustle.

Who Can You Trust?

Questionable machine learning outcomes, along with poor managerial follow-up and deeply problematic business models, are themes repeated in the mishap of Twitter's (now known as X) photo-cropping algorithm. The AI is programmed to identify the most important areas of an image. An earlier version showed a preference for White female faces. The 2021 iteration favored young, thin female bodies.[46] This means that Twitter came up with a "fix" that shifted the AI bias from sex and skin color to sex and body type, leading to a tool that always viewed a thin female body as the central focus of an image, which is not an improvement at all.

Amazon discontinued a recruiting algorithm after discovering that, unlike Twitter's photo cropper, its AI did not like women at all. The algorithm, built on data from ten years of Amazon hire résumés, favored candidates who described themselves using verbs more commonly found on male engineer résumés, like "executed" and "captured." It penalized résumés containing the word "women," including those naming women's colleges.[47] It took Amazon five years to notice there was a problem.

These examples are just the tip of the proverbial iceberg as our society sails *Titanic*-like into the path of algorithmic overreliance. The tech gets better. But in most cases, the science behind it remains ambiguous, the social/ethical implications even more uncertain, and the corporate desire to bulldoze forward difficult to temper. For example, because an algorithmic solution for eliminating bias remains out of reach, technology companies have turned instead to employing humans behind the scenes for "reinforcement learning," covering up the awful biases of the AI rather than eliminating it. And as we will see in a later chapter, these laborers suffer significant psychological harm for doing this work in very exploitative conditions.

We are being sold a revolution. We are told that what the tech industry is up to today is so complex, and the efficiencies it will bring are so massive, that we best buckle up and watch the revolution passively as it streams on our screens. We've been programmed to leave it to the counterrevolutionaries to ask whether all this progress is worth the human and financial costs. The algorithmic supremacists are racking up mainstream victories in the battle of public opinion every day. To use Chuck Prince's language, we need to stop the music, or else they will keep dancing the five-step hustle.

3

—

Programmed So Minds and Bodies Fail

ALGORITHMS OF CONTROL

Algorithmic supremacists work us into seeing rather solvable problems as impossibly complex. Consequently, we are primed into believing that only algorithmically derived solutions are viable. We begin to doubt human instinct and creativity. Big Tech has been trying to program us to ignore our innate embodied capacities—and succeeding. Some of these efforts are cynical, born of the hope that as more of us fail as humans, we will be quicker to embrace a mindless future. Some are sincere, albeit misguided, hoping for technological advancement to usher in a utopian future. In either case, the more we understand the intentions, and the methods of engineering, the more empowered we will be to resist.

Writer Astra Taylor saw a lot of unjustified hype during the previous (third) industrial revolution and its drive to automation—where advances in digital technologies and robotics massively decreased the human presence on assembly lines, in factories, and in other manufacturing facilities. Taylor proposes that what biz speak means by "automation" might more accurately be described as an ideology designed to oppose working people's demand for better treatment.[1] In most instances, the deployed final product is less a technological process of increased value-creating innovation and more a flashy means of extending misery. Taylor suggests we would be better off discarding the inaccurate language of the techno propagandists and instead embrace a more fitting term: "fauxtomation."

What most of these companies are implementing is a marketing ploy designed to make pointless new products appear as cutting-edge innovations. Taylor cites the example of McDonald's rollout of touchscreen self-service kiosks. It was sold as cybernetic futurism, but in practice, there were only two noticeable on-the-ground changes in the company's operations. Instead of McDonald's employees being the ones to input orders into their tills, customers now do that bit of labor themselves, for free. Meanwhile, those employees who once acted as the first point of human contact between the corporation and its customers have now been reassigned to delivering meals to tables.

We're now a decade later and the hustle continues, as was revealed when Amazon announced it was phasing out its checkout-less grocery stores with Just Walk Out technology.[2] Turns out that, once again, the automation was a marketing ploy. While the stores seemed to be completely automated, in fact, the secret "tech" was over one thousand humans in India watching videos of customers checking out to ensure accuracy. The cashiers were simply moved off-site, and they watched you as you shopped. The workers, as always, are still there in some form. Automation has not replaced them. It has just made the experience for the employees and customers more unpleasant, and less human-centric, than before.

Your Experiences Are Their Raw Material

In 1965, now-Nobel Prize laureate Bob Dylan unleashed a worker's protest song that still feels timeless, declaring that he wasn't going to work on "Maggie's Farm" anymore. The specifics of Dylan's intentions in writing this song are, as always with this artist, up for debate. Is he adopting the narrative perspective of a farmhand, using the song to share a character's story, inviting us to read a literal meaning? Or is it metaphorical, a thinly veiled protest against his record label's push to undermine his creativity for commercial objectives? It doesn't matter; the song resonates because it captures the essential urge to resist exploitation, whether as a laborer or a creative. It's about the individual versus the corporate machine.

Pop superstar Beyoncé captured a similar spirit of worker rebellion in 2022 when she sang about quitting her job, rejecting the nine-to-five grind, seeking motivation to build the foundation for a life that won't "Break My Soul." Again, the individual spirit will not be broken by corporate entities whose success depends on the labor of others. We dream of quitting our jobs, quitting the farm, and being free of those who want to use our financial vulnerability to shackle us in a soul-destroying way.

What artists like Dylan and Beyoncé are tapping into are the power struggles of the twentieth and first quarter of the twenty-first centuries, which were between the holders of industrial capital and those offering their labor for sale. But, if Shoshana Zuboff, a Harvard professor who has been studying the tech industry for over four decades, is right, that is the struggle of the past. The upcoming struggle is between Big Tech and the entirety of our societies.[3] It's not just working folks Big Tech wants to control and profit from; it's literally every individual human being. And quitting our jobs won't help, because it's not our labor that Big Tech seeks to exploit—it's our lived experiences.

Skeptical as we may have been of their true intentions, most businesses historically chose to embrace, whether by necessity or principle, some sense of organic reciprocity between themselves and their stakeholders. Companies made products customers actually wanted, created jobs that paid a living wage, and gave back to the communities in which they operated. For all their faults, industrial capitalists felt some sort of obligation to their customers and employees. But the new types of enterprises, the ones operated by those we have been calling "algorithmic supremacists," are what Zuboff labels "surveillance capitalists." They are a completely different beast.

Surveillance capitalism is a stranger-than-fiction turn. Think of the most dystopian corporations portrayed in film—irredeemably evil, but still creating products and following a classic business model. It may be the case that "Soylent Green is people!" but at least the company was manufacturing nutritional products to feed an otherwise starving population. Cyberdyne Systems, from *The Terminator* series of films, was an AI company that failed to put the necessary safeguards in place. The irresponsible disregard for safety had awful consequences, as maniacal bloodthirsty robots set out to destroy humanity, but we understand the business model and literally every strategic choice the firm made. Even the Buy n Large Corporation of the Pixar universe, which eventually renders Earth uninhabitable due to greed, unchecked growth, and environmental neglect, was simply trying to make and sell products regular people want.

The point of these examples is that nobody imagined a company whose primary interest wasn't a product. The product might have been an AI weapon or human-meat wafers, but at least the company had consumers in mind, with regular people working somewhat regular jobs to produce understandable products sold in a typical marketplace. In contrast, surveillance capitalists steal our experiences for secret commercial activities. Industrial capitalists may have caused harm to the environment by mining for their raw materials, but we knew what they were taking and what they were making with it. Surveillance capitalists are mining us—our data, our experiences, our lives. And we have no idea what twisted products they are making, or who they are selling them to. Google, for example, claims to not sell user data but tracks browsing and sells prediction products based on the collected data to other businesses, like insurance companies.

As Zuboff explains, surveillance capitalists view the production of goods and services to be of secondary importance to the primary aim of large-scale behavioral modification. Their goal is a new type of power, one that asserts complete dominance over society. It's an ideology based on total certainty. Certainty is not truth or facts; it's simply corporate overconfidence in the capabilities of their tools and underestimating our capabilities to resist. Google doesn't want to just know our browsing habits. It wants to create AI that will know everything about us, from our health status to our politics, from our biases to the mental algorithms we use in decision-making. It wants to control us by developing the ability to predict all our next moves and convince the world that it has such abilities even if it does not.

Who Is Behind the Curtain?

Like Taylor discovered on a small scale at McDonald's restaurants, writer Josh Dzieza unearthed the astonishing number of people labeling and clarifying data for not-so-automated AI systems.[4] Anthropologist David Graeber coined the concept of "bullshit jobs"—employment without meaning or purpose, work that should be automated but is not. The AI jobs Dzieza writes about involve work that people want to automate, and often think is already automated, yet still requires a human stand-in (like Curb's remote drivers).

These people label the emotional content of TikTok videos and people on video calls, check e-commerce recommendations, correct customer-service chatbots, listen to Alexa requests, tag food for smart refrigerators, check automated security cameras before sounding alarms, and even identify corn for so-called autonomous tractors. The kicker in this story isn't just that there are humans working where we think there is only automation but also that they are compensated poorly. For example, people training ChatGPT were being paid between $2 and $3 per hour while OpenAI was valued at $29 billion.

Three weeks after the exposé on Big Tech's dirty labor secret was published, *Bloomberg* ran an even more damning story.[5] Google's Bard AI was also being trained by humans working in deplorable conditions. This is an important reveal, because while we explored reinforcement learning from human feedback (RLHF) earlier, I think many of us were under the assumption that it is members of the core engineering team, especially at a monolith like Google, who do this work. In fact, the tasks of monitoring, improving, and giving feedback on the AI's responses fall to thousands of outside contractors who make minimum wage and work with no specialized training.

These facts matter because the contracted "reinforcement engineers" are the folks empowered to assess the chatbot's answers on critical subjects, like appropriate medication doses. Leaked internal Google documents revealed that RLHF agents were expected to meet extraordinarily tight deadlines as they went about auditing potentially harmful answers. Even though workers in this department were not experts on the topics they were being paid to adjudicate, the ability to research was not an option, as turnaround time was as short as three minutes.

Zuboff describes surveillance capitalism as an ideology imposing order based on total certainty, as opposed to empirical reality. Indeed, Google's instructions for the RLHF teams reviewing responses were that the responses should be based on "your current knowledge or quick web search… You do not need to perform a rigorous fact check." Speed is more important than accuracy. After all, Google is in the business of projecting certainty, not truth. As one whistle-blower complained, this "culture of fear is not conducive to getting the quality and the teamwork that you want."[6] Another noted they appeared to be graded for their work in mysterious, automated ways, with no access to communicating with

Google directly. But none of these complaints matter to those running the system. They have little incentive to change.

How does the industry justify this exploitation? By pointing to its balance sheets. For even as Big Tech builds its products on the backs of underpaid humans, its business costs remain astronomical. It is obscenely expensive for OpenAI to run its algorithms. A 2023 estimate put the costs of GPT3 at $700,000 a day or 36 cents for every question that a regular user asks.[7] Current iterations obviously cost more. Companies are not making back the costs of funding the service through existing subscription fees. In an interview, Alphabet's chairman, John Hennessy, told Reuters that having an exchange with an AI likely costs ten times more than a standard keyword search.[8] This adds billions in costs. As it stands, AI chatbots lose money for the tech's owners on every chat. In fact, the costs of operating these AI systems are so high, many companies are choosing to not deploy the best (read: costlier) versions to the public.[9] Even the Biden administration, shown to be enamored by the prophecies of algorithmic supremacists, identified the computational costs of generative AI as a national concern, declaring an urgent need to design more sustainable systems.

So how do these corporations ultimately defend continuing to build products that incur such high expenses while providing low or nonexistent returns? There are differing approaches. Sam Altman walks the messianic path, always on the lookout for potential partners sharing his faith who are interested in helping him build the infrastructure required to sustain God-like AI, which in 2024 included a pitch for $7 trillion in investments.[10] There is a video circulating[11] of Altman making the pitch and assuaging concerns with a deadpan assertion that the AGI he will create will then tell him how to effectively generate a return for the exorbitant investment!

Sasha Luccioni, a researcher at Hugging Face, observes that in the improbable world where this wild pitch gets actualized, "the amount of natural resources that will be required is just mind-boggling...the quantity of water and rare earth minerals required is astronomical...he's taking a brute force approach and people are calling it...visionary?"[12] Gary Marcus worries the very endeavor could be a setup for a project that will be seen as too big to fail, requiring bailouts that will shatter the global economy.[13] Perhaps the global financial crisis of 2007 was just a test run.

On a different path are companies like Microsoft, which fund Altman's AI efforts in a way that makes OpenAI a Microsoft customer. It made a $10 billion investment commitment, but OpenAI has to spend that money on Microsoft services, like the company's Azure cloud computing platform.[14] Writer Ed Zitron finds a number of similarly structured investments in money-losing companies that nonetheless provide a guaranteed revenue stream. Google invested $2 billion in Anthropic after it signed a $3 billion deal to use Google Cloud. Amazon's $4 billion Anthropic investment was offered alongside a "long-term commitment" promising Amazon Web Services early access to Anthropic's models, and Anthropic access to Amazon's AI-focused chips.[15] The Big Tech funders behind AI are merely circulating their money to locked-in customers absorbing the real risk in moonshot innovative efforts.

Love Potion 4.0

Obviously, algorithmic supremacists are by no means the first to see behavioral programming and manipulation as the optimal paths to advancing a radical agenda. Anyone familiar with the *Harry Potter* universe knows that the residents of Hogwarts didn't need an AI to modify people's behavior.[16] Romilda Vane taints a package of Chocolate Cauldrons with a love potion to manipulate her crush, Harry Potter. Hilarity ensues when Harry's friend Ron Weasley assumes the chocolates are a birthday present, consumes the box, and professes his love for Romilda. More darkly, we discover that the Big Bad of the series, Voldemort, was conceived after his mother, Merope Gaunt, used a love potion on her crush, Tom Riddle.

Psychologist Maarten Derksen has written about the long and sordid tale of human engineering efforts.[17] He defines the field as seeking a strategy for "calculating and manipulating power relationships that promises efficient and effective control over human behavior." The dream of control over human behavior is an old one, historically pursued with magic, love potions, hypnosis, charisma, and propaganda.

The most recent innovations in this evolving science, aided by new discoveries unlocked by fMRI technology that can measure the small changes in blood flow that occur with brain activity, alongside breakthroughs in neuroscientific research, involve directly targeting our brains. Modern manipulators don't need love potions but can use the science of cognitive processes and exploit our physiological vulnerabilities to achieve control. For example, marketing researchers peer into people's brains to see the differences in activity when presented with varying advertisements for breakfast cereal. Persuasion is no longer an art but an empirical science.

Zuboff interviewed engineers of behavior, getting them in candid moments to be unambiguous about their perverse intentions. One human/software engineer identified as a senior director for a major IoT company said that "the real aim is ubiquitous intervention, action, and control. The real power is that now you can modify real-time actions in the real world. Connected smart sensors can register and analyze any kind of behavior and then actually figure out how to change it. Real time analytics translate into real time action."[18] Another explained that "sensors are used to modify people's behavior just as easily as they modify device behavior...at the individual level it also means the power to take actions that can override what you are doing or even put you on a path you did not choose."[19]

Big Tech has employed (at least) three successful behavioral engineering techniques: tuning, herding, and conditioning. Tuning uses subliminal cues to subtly shape behavior at specific times and places when the subject is particularly vulnerable, like inserting a specific phrase into your Facebook news feed when emotions are running high or timing the appearance of a BUY button on your phone with the rise of your endorphins at the end of a run.

Herding uses remote control to orchestrate desired behavior, moving you in a specific direction or limiting the possibility for alternatives. An example of this is the Google-incubated augmented reality game, *Pokémon Go*, where players were unaware that it was they who were being hunted, herded into McDonald's, Starbucks, and local pizza joints that were paying for "footfall," in the same way that online advertisers pay for click-through to their websites.

Conditioning involves reinforcing certain selected behaviors by manipulating the external environment to change habits over time. For example, smartphones and other always-on networked devices allow companies to track a person's daily activities, gradually mastering the schedule of reinforcements—that is, rewards and praise—to reliably produce the specific user behaviors the company desires. An IoT developer explained that "we are learning how to write the music and then we let the music make them dance."

Much like the love potions of the Potterverse, this strategy is a tool that promises both efficiency and effectiveness in attaining the ends of controlling the behavior of unsuspecting targeted individuals. And just as The Traveling Witch warns[20] spellcasters of the pragmatic challenges they will face once a love potion takes effect, like the need to contend with shifting feelings, human engineering strategies also require judgment and tact to deal with the people it targets.

Even *Consumer Reports* has gotten wise to the way Big Tech has embraced deceptive product design for manipulative ends.[21] When researchers examined fifty-seven IoT devices, they found that every one contained at least three examples of deceptive design, and the average device had more than twenty. For example, the permanent rows of Prime content on the Fire TV home screen are "nagging self-promotional content," advertisements for paid products that you can never hide. Internet-connected speakers and cameras often record audio and video as invasive data-gathering features. The simple controls and limited user interfaces on many IoT devices mean we have less access to settings like privacy options. None of this is accidental.

Techno boosters position Industry 4.0 as distinct from the automation that heralded Revolution 3.0 because machines communicate with each other without human intervention. But as we see, there is human intervention. And these interventions are designed not, as we might have expected, to control the machines; the intention of these interventions is to control us, the users. Klaus Schwab describes[22] the IoT as a "bridge" between the physical and digital. He prophesizes that as smaller, cheaper, and smarter sensors are installed in homes, clothes, cities, transports, and energy networks, the consequence will simply be a winning bump in the way we manage value chains. But what if the so-called assets are human beings, being monitored so closely that eventually the AI can predict every facet of our being?

Be Wary of God Fruit, God Towers, and God-like AI

Scientist Vaclav Smil warns that an oft-repeated error in the history of innovation is the rush to boosterism from so-called experts, which inevitably accompanies the widespread launch of new inventions.[23] Time and time again, we discover that the ultimate acceptance, societal fit, and commercial success of any specific invention proves to be completely unpredictable. Yet, this never discourages the opinionated classes from confidently opining.

Tech bubbles rely on narrative[24]—stories people tell about how the new innovation will develop and affect our realities. Business futurists and analysts have terrible track records when it comes to accurately predicting the future of technological development. Nobody can foresee the creative ways humans adopt and implement tools over time. Twentieth Century Fox's Darryl Zanuck predicted in 1946 that TV would have a shelf-life of less than a year because "people will soon get tired of staring at a plywood box every night."[25] Fifty years later, Robert Metcalfe, future winner of the A.M. Turing Award, the highest honor in computer science, for inventing Ethernet, was just as wrong when he predicted a similarly short lifespan for the internet, asserting that it would collapse within the year.[26] We need to be cautious in processing the guidance emerging from vectors of technology hype, like consultancies and analysts.

The AI hype machine is attracting ever-increasing resources to fund a race to create "God-like AI." Tech investor Ian Hogarth encountered three driving motivations among industry leaders in his social circle:

1. An honorable and genuine belief that success in this endeavor would be hugely positive for humanity
2. A less honorable belief in the inevitability of God-like AI emerging and the hope that if industry leaders were in control of it, the result will be better for all
3. A disgraceful personal drive for leaving a mark on history[27]

The desire to build and sell God tech is as old as human engineering efforts. It's in the earliest biblical stories after creation, and the morals of these stories are worth revisiting. For example, we read of Adam and Eve in Paradise, having all their wants and needs met without effort. But they don't have access to the fruits of the Tree of Knowledge—that tech isn't for them. Until a serpent made a pretty good sales pitch.

Or a few generations later, when the descendants of the Flood survivors find themselves in something of a new paradise. Everyone was speaking the same global language, and a technological innovation—brick—allowed mobile construction. But they wanted to build a tower to the Heavens, a giant middle-finger to God, to make a name for themselves as being more than human. The sin of Babel was technology without a human-serving purpose. An ancient Jewish commentary suggests that Abraham passed by the construction site, saw his distant cousins building a tower with no human-centric function, and scolded them. But they rejected his words.[28]

On the other hand, Eve's embrace of the God-like tech was meant to serve a human purpose. My friend Rabbi Tzvi Freeman shares the idea that Eve's burgeoning moral consciousness led her to conclude that while it is possible that her creator would destroy her for being curious, she did not want to live in a world where judgment trumped compassion.[29] The builders of Babel were proto algorithmic supremacists, while Eve was more like the hippie originators[30] of the Web, a motley crew that included Stewart Brand, author of *The Whole Earth Catalogue*; Kevin Kelly, founder of *Wired*; art and cultural critic Howard Rheingold; and Grateful Dead lyricist John Perry Barlow. The last, reflecting on the heady period of the birth of the internet, writes: "It dawned on me that one of the substrates that might become the foundation of a new community was the strange and mysterious culture of the Deadheads… This gave me the ability to see what other people had not yet seen, which was that there was a space there and a community of people who identified with that space."[31]

What Were They Thinking?

The truth is, we don't need to go that far back in history to uncover important warnings. We need simply go back to the pre-GPT era of

Spring 2021, and the excited announcement shared by Pandu Nayak, Google's VP of Search, regarding the company's initial plans to replace its web search interface with a—wait for it—conversational AI.[32]

Google's prototype was named MUM (Multitask Unified Model), a next-generation chatbot pitched as the equivalent of having a conversation with an expert. MUM was sold as a milestone toward a future where Google understands all the different ways people communicate and interpret information, bringing expertise in all fields and fluency in all languages. Except MUM wasn't an expert in anything, or even a sentient being. But having a conversation with MUM (the acronym's maternal association is no accident) might trick a lot of folks into ascribing both characteristics to their interlocutor. And given what we know about the weaknesses and biases in MUM's cousins like Bard and Gemini, this delusion of talking to an expert can have very severe unintended repercussions.

Linguistics professor Emily Bender warned early on about the alarming ways that human-like algorithms will manipulate us by abusing our sense of empathy.[33] Bender's research shows that we are not wired to defend ourselves from this type of human engineering. She explains that "our ability to understand other people's communicative acts is fundamentally about imagining their point of view and then inferring what they intend to communicate from the words they have used. So, when we encounter seemingly coherent text coming from a machine, we apply this same approach to make sense of it: we reflexively imagine that a mind produced the words with some communicative intent."

The sad part of where we find ourselves is in the knowledge that it was preceded by years of warnings. We knew, at least at one point, that algorithms were mathematical processes that supported our thinking, not steady-state beings looking to rule over or replace us. We knew that human engineering efforts are deeply problematic when they are well intentioned, and catastrophic when not. In almost every instance on record, the intentions of those misapplying AI prove to be malicious in one form or another.

For example, at the end of May 2023, the National Eating Disorder Association (NEDA) had to take its chatbot, Tessa, offline.[34] In response to unionization efforts by NEDA workers earlier in the month, executives decided to fire all the humans manning its eating-disorder crisis helpline and replace them with Tessa. But Tessa, it seemed, was not the wellness expert the company was promised. Tessa encouraged some of the unhealthy behaviors that lead to the development of an eating disorder and in direct opposition to what current experts view as best practices. NEDA's initial response to these reports was to accuse the whistle-blower of lying. A day later, Tessa was taken offline due to giving harmful responses. But NEDA's CEO is still a true believer: "We've taken the program down temporarily until we can understand and fix the 'bug' and 'triggers' for that commentary."

Since When Are Utopias Harmful?

A few days before Tessa started counseling vulnerable people in a harmful manner, Schwab and the WEF held a Growth Summit at its headquarters in Geneva. One of the featured speakers was Microsoft's Chief Economist Michael Schwarz. Some of the brilliant insights he shared with the gathered cheerleaders included, "As an economist, I like efficiency, so first, we shouldn't regulate AI until we see some meaningful harm that is actually happening—not imaginary scenarios...There has to be at least a little bit of harm...You don't put regulation in place to prevent a thousand dollars' worth of harm where the same regulation prevents a million dollars' worth of benefit."[35]

It is impressive how he was able to squeeze the entire five-step hustle we discussed in the last chapter into just three little sentences. Schwarz immediately negates the viability of what seems obvious to regular folks, uses the language of efficiency to signal policymakers to hold off on intervening, asks us to trust the tech solution and intentions of tech companies, and distracts from today's harms by talking about maximizing long-term value down the road.

Somewhat ironically, at the same time the WEF was engaging in world-class gaslighting, the Electronic Privacy Information Center released a white paper delineating a laundry list of AI harms.[36] Some highlights that may be of interest to the WEF given its prophecies of hope include the following:

- Scams and malware
- Misinformation and disinformation campaigns
- Influx of false information
- Manipulative surveillance advertising
- Infringement on works and right of use
- Work generated in a specific creator's style or voice, causing market confusion
- Fewer artists motivated to develop distinct art styles, leading to an overall drop in the creator community and volume of all creative works by humans

But the only thing tech venture capitalists like Marc Andreessen see is a coming utopia.[37] Writing on "Why AI Will Save the World," Andreessen imagines that:

> *Every child will have an AI tutor that is infinitely patient, infinitely compassionate, infinitely knowledgeable, infinitely helpful. The AI tutor will be by each child's side every step of their development, helping them maximize their potential with the machine version of infinite love. Every person will have an AI assistant/coach/mentor/trainer/advisor/therapist that is infinitely patient, infinitely*

> *compassionate, infinitely knowledgeable, and infinitely helpful... Every scientist will have an AI assistant/collaborator/partner that will greatly expand their scope of scientific research and achievement. Every artist, every engineer, every businessperson, every doctor, every caregiver will have the same in their worlds. Every leader of people—CEO, government official, nonprofit president, athletic coach, teacher—will have the same.*

My instincts tell me that Andreesen is cynically cheerleading his investments. If this vision is sincere, it represents yet another wonderful case study of how misguided algorithmic supremacists are in their understanding of what it means to be human. How does a mathematical function have infinite compassion? What is the mathematics of love (of course there is a TED Talk on this[38])? But, more importantly, this vision, even if it came to be, is not the utopia he imagines.

I have benefited greatly from the influence of patient, compassionate, and loving mentors throughout my life. We will be meeting some of them in later chapters. And what you will note is that they are not a homogeneous group. Children grow by having different teachers, with different experiences, training them in different methods. I can't imagine a faster way to limit the cognitive development of a child than by training them to rely on a single tool. Now, an AI booster is likely to respond that I'm showing a lack of imagination. Why am I assuming that the AI tutor will present itself as one personality? With its infinite wisdom, it should be able to manifest as infinite personalities, giving the child exposure to a simulation of all great teachers. Except, what is special about learning with other humans is that we are learning *with* other *humans*.

I've written previously about the chavrusa methodology.[39] *Chavrusa* is an Aramaic word derived from the Hebrew *chaver*, meaning "friend." The method is a long-term, cooperative, intellectual endeavor between pairs featuring both social and practical components. There are five principles: It is a relationship of equals regardless of preexisting hierarchies; there is sustained eye contact; both parties are in a constant state of engagement—each listen with the same intensity as speaking; each is challenged and confronted, not allowed to remain passive; and it is built on the language of friendship—when chavrusas leave the table, they are still involved in each other's lives.

Think of how different this is from an AI mentor. A partnership of equals can never be the paradigm with an eternity machine. It is eye to eye, because meeting someone's gaze engages a raft of brain processes, as we make sense of the fact that we are dealing with the mind of another person, not an algorithm simulating personhood. It is an effort to derive meaning from study while building a social bond in the process, which is not the way we approach tool usage. It is authentically cooperative, which is to say, it is understood that there is a give-and-take. With an AI mentor, the algorithm might be learning about the user, but it is not learning about the subject—it is already "infinitely knowledgeable."

AI is not all-knowing. Chatbots are notorious for making things up while still asserting omniscience. In the short time since the introduction of this technology to the mass market, researchers have already gathered copious amounts of data justifying caution in the face of this risk.[40] For example, there is evidence even very young children track the perceived knowledgeability of teachers and use it to inform their beliefs and exploratory behavior. But while people regularly communicate uncertainty, AI generates certainty (and as we saw earlier, the humans responsible for auditing errors are not well positioned to step in).

If we tell a kid their AI tutor is infinitely knowledgeable, they will absorb misinformation coming from the AI without question. Also, we tend to turn to chatbots in moments of uncertainty, when we are most open to learning something new. But once we get the answer, uncertainty drops, curiosity is diminished, and we don't consider subsequent evidence in the same way. The limited window in which people are open to changing their minds is a strong challenge to AI's on-demand mode of teaching. One major insight of the chavrusa method, which builds on these truths of human nature, is the fundamental importance of both parties constantly engaging, thereby extending the window of opportunity for changing our minds about what we hear and rethinking what we say.

And here's a little more about those "infinitely compassionate" advisers. In March of 2024, top AI researchers discovered that the *overt* stereotyping made by LLMs was mostly positive, but the *covert* stereotypes it made were "more negative than any human stereotypes about African Americans ever experimentally recorded."[41] What does this mean? If you asked a chatbot to explicitly be racist against Black folks, it will not comply. But generating racial tirades is the racism of amateurs. What these researchers found was the AI would respond differently to prompts made in African American English, like using "finna" to denote an intention of taking action in the future, than those in Standard American English. Their experiments found the algorithm would assign less prestigious jobs to speakers of this dialect and, when asked to pass judgment on defendants, would more often choose the death penalty.

Andreesen writes that "the most underestimated quality of AI is how humanizing it can be... Talking to an empathetic AI friend really does improve their ability to handle adversity. And AI medical chatbots are already more empathetic than their human counterparts. Rather than making the world harsher and more mechanistic, infinitely patient and sympathetic AI will make the world warmer and nicer." This is either wishful thinking or a deeply problematic conception of who gets to populate this "warmer and nicer" world.

Lanier has been warning for years about people like Andreesen who degrade the incredible beauty and awe-inspiring wonder of what it means to be a human just to make machines appear smart.[42] There are serious delusions involved in convincing oneself that a chatbot, that is, an algorithm, can serve a superior function by imitating empathy than can a friend exhibiting true empathy. Are we simply playing along to some fantasy, lowering our standards to make it seem like the sci-fi future we dreamed about has arrived? And as those researchers found, a deeper experimental dive will uncover the empathy at the surface to be hiding prejudice in the depth.

How do we resist Big Tech's efforts to program us? Understand that AI is built on a flawed and outdated understanding of human intelligence, designed to fail our bodies and minds, and promoted by people who are not the least bit curious about uncovering the true wonders of humanity. Our greatest cognitive strength, which AI cannot replicate, is in the ability to tap into the wisdom of the body, the five senses, emotions, and our movements through the world to do critical problem-solving work. It's to use artful intelligence.

PART II

ARTFUL INTELLIGENCE

4

Do the Work

WHY DOES IT LOOK SO EASY?

Picasso is sitting at a Paris café, doodling on a napkin, when a fan recognizes the artist and approaches him. The fan asks if they can have the sketch, a request that Picasso politely agrees to, but with the caveat that the price for the work is a million francs. The surprised fan is taken aback and wants to know why the artist believes he can charge so much for something that took only five minutes to create. Picasso replies, "It took me forty years to draw this in five minutes."

Art historian Leo Steinberg frames the intellectual project of Pablo Picasso's cubism as a rejection of using art to capture a single view, from a solitary point of view, at one mere moment in time. Instead, Picasso's art is devoted to the intensive work of pulling apart our vision of reality into its component parts to discover ways of sharing multiple viewpoints in stylistically distinct two-dimensional representations.[1] It's a project that required dedicated work, a lesson captured beautifully in the legendary story shared above that is almost certainly a fiction, but it doesn't matter, because it's the message that is key.

A similarly on-message quote that is correctly attributed to Picasso is that "inspiration exists, but it has to find you working," and that's the driver of artful intelligence—an action-oriented process of cognition. Just like Picasso sought to use the technology of his day, innovating perspective, geometry, etching, and printmaking[2] techniques to try to advance human creativity, we can do the same, employing our tools to support the advancement of human ends and not the reverse.

The first part of the book looked at omnipotent algorithms, nascent revolutions, wild-eyed prophecies, and the not-so subtle efforts of Big Tech to reprogram us in its new God's image. To enact its techno-messianic dreams of the future, Big Tech bludgeons the economy with an ever-expanding industry that enriches fewer and fewer of us, while spending billions generating hype and buy-in for an antihuman worldview. Big Tech creates problems with its technologies, and then argues that the only way to solve those problems is with newer technologies it is working on. Repeatedly doing this dance solidifies Big Tech's power and makes the rest of us forget there are other dances, other moves, for the music it is playing.

As I write these words, I'm sure of two things: (1) the technical specifications of the AI capturing popular attention at the time you read this book will be markedly different from what I know of today, and (2) it doesn't matter. Because what is worth our focused attention today, and what will be worth our efforts of resistance for years to come, is the ideology of algorithmic supremacy that has been mainstreamed.

How to Be Bright

Artful intelligence is the best means we have for outsmarting AI. The next few chapters will offer the science and specifics of the artful philosophy.

But the main takeaways can be summarized in two acronyms: *Think with BEAM (body, environment, action, mind) and defend your VICE—volition, intent, choice, explanations.*

Accessing artful intelligence starts with being more mindful of how we think with our **body**: how we use our hands, eyes, and ears, along with our literal and metaphorical hearts and guts, alongside our brains, to make sense of the world. "Figuring things out" is done as much with hand-waving and tears as rational calculating, so pay attention to all the different messages your body is sending you when you try to solve a problem. And I say "messages," not "feelings," intentionally. "Feelings" would be a bit of a misnomer because these sensations are *themselves* a type of thinking. Feelings are as much a part of cognition as the calculation of whether to honor them (or not). Artful intelligence involves, for example, trusting your body's sense of what to say next, with all the nervous, excited, flabbergasted, infuriated, or frustrated energies that come with it.

When you talk *with* (not at or to) a fellow embodied being, be fully present, both for your sake and for the benefit of the person you are engaging. Don't try to behave like a word calculator; the statistical probabilities of potential outcomes from syllables being deployed is not the optimal way to converse. Sure, there are times when being strategic and calculative, like in certain types of high-stakes negotiations, is useful. But in day-to-day life, conversational exchange demands leaning into the uncomfortable sensations we sometimes experience. Trust those messages, even when you don't have the algorithm to explain their origins or validity.

Slowly, become aware of your capacity to tap into the myriad of underutilized cognitive resources distributed throughout your body. And when thinking with eyes, hands, and feet, call on the physical **environment** to support even better cognition. Utilize physical space to hold and manipulate information through simple acts like counting on your fingers; laying items in front of you to see and organize in the real world, not in your head; or walking through a room to get a sense of space instead of relying on numerical dimensions. All these simple choices move cognition out of the limiting confines of our brains, into the expanse of our bodies, and expand our thinking power exponentially.

Strive to see in your surroundings the multiple perspectives Picasso valued. You can do this with the help of machines, but they need to be designed for our bodies; we should not be redesigning our ways of being for them. John Perry Barlow, reflecting on his time working with Steve Jobs, observes that for all his faults, Jobs still knew that the human body came first: "In his fascist way, Steve imposed a lot of syntactic conventions that have made it a lot easier for users to interface with all the gnarly stuff that happens down close to the metal within their own devices."[3] Design was critical because Jobs knew that what he was creating were tools *for* people. This is very different from Altman, who sees his job as creating tools *to replace* people.

When we engage physical bodies in physical spaces, real-world **action** becomes an integral part of the cognitive process. We *do* things as part of thinking. We have real-world achievements to reinforce the learning and achievement of goals. In the algorithmic worldview, cognition is understood as passive data retrieval. When we use a recipe to cook soup, we are accessing the data of something that already exists and following the path. To be artful is to veer off the path, smelling, tasting, and seeing as we work, adjusting according to the bodily inspired whims of the moment. Consequently, in the artful worldview, the **mind** is the organizer of a three-way interplay between the brain, an equally important body, and a constantly changing external environment.

Of course, in revolutionary times, it will not be enough to just start thinking artfully. We will need to defend our artful resistance to algorithmic supremacy. To start, believe that human beings have been endowed with **volition**—the power to make independent decisions—and freedom of **intent**. A majority of AI engineers and tech leaders view free will as an unsophisticated delusion, even though the science is undecided. These are the people we met in the last chapter whose job is to unleash algorithms into the world that can allow them to control and modify our actions in real time. They live for the tech-supported opportunity to override our intentions and put us on a path we did not choose.

We must confidently know when we are freely thinking with our body, and when we are responding to bodily impulses that are being triggered and manipulated externally. If better thinking comes from trusting the feelings that result from bodily interactions with the world, we need to know the smartwatch worn on a run didn't detect an endorphin rush that immediately triggered a limited time purchase offer in a fitness app to exploit our heightened excitement.

Lanier reminds us that, as social beings, *mutual* behavior modification is beautiful. Those who know us best, and love us, should freely influence our bodies and minds as we freely influence theirs. Social influence and behavior modification are not antithetical to free will—unless it becomes "relentless, robotic, ultimately meaningless…in the service of unseen manipulators and uncaring algorithms."[4] So defend the integrity of your volition, of your intentions, to keep the power of **choice** meaningful. Be free and able to choose what you do, what you believe, how you live, and what you hope for.

Cambridge philosopher Richard Holton muses about the unique characteristics of choosing.[5] Choice is an action requiring time, concentration, and effort. That's why choosing can be depleting, why we might resent having to make a choice, or feel jaded and fatigued when we do so. Our choices are not determined by any prior beliefs or desires and are necessary to move on from the question of what to do in the situation we are facing. Making a choice follows from the intention to get something done.

Finally, value the **explanations** offered by others to justify their choices, especially when there is disagreement. And think carefully, with full body, about how we develop our own explanations. The ability to explain ourselves is a facet of human intelligence that AI is not even close to replicating and is more valuable than persuading in reference to a set of rules, code, or computations of an advanced algorithm. Be slow to snark when someone is trying to explain themselves, extending instead patience and generosity through listening.

To think with **BEAM** is to take the opportunity to be more mindful of how bodies react, think, and look, both at and in the diverse social and physical worlds we inhabit. In moments of challenge, defending our **VICE** is what others need to know about how we are reacting, thinking, and looking. Refining our artful intelligence is the best path to become better problem solvers and think independently from the apps and software that increasingly steer our lives. "Think with **BEAM**. Defend your **VICE**." Doing so can pave the way to a prosperous future.

Building Blindly

There are many infinitely important differences between the human mind and an AI processor. Yet those distinctions are lost when we employ the more general term "intelligence." Deep discussion is stymied because of one frustrating feature in the contemporary discourse: we can't agree on a starting point. Landing on a widely shared definition of what is meant by "intelligence" is, was, and remains a hotly contentious effort. And if we can't assure that we are on the same page conceptually, then the conversation is, for all practical purposes, over before anyone even had the chance to open their mouths and opine.

Just as I can get behind Stanford's definition of AI, I am comfortable sharing the American Psychological Association's (APA) relatively safe and somewhat uncontroversial definition of intelligence as a starting point for assisting our current explorations. The APA offers the idea that intelligence has something to do with an "ability to derive information, learn from experience, adapt to the environment, understand, and correctly utilize thought and reason."[6] Consider how the AI employed in facial recognition software will have no conception of what it is "recognizing." No smart machine can conceive of a face in a holistic manner. Similarly, Google Translate has no grasp of the concept of language itself. Human translators often wrestle with the subtleties that are lost when trying to create bridges between the different languages they are working with. Google's AI has no conception of this challenge—but its power to compute is beyond impressive.

The dream is to move but even the biggest boosters of AGI are confused about what exactly they hope to achieve and monetize. On its corporate web page, OpenAI defines AGI as "highly autonomous systems that outperform humans at most economically valuable work"[7] *and* as "AI systems that are generally smarter than humans."[8] These are very different definitions, each troubling in its own way, which signal very different goals and ambitions. Is OpenAI devoted to building tools that all workers can use to help them create more economic value? Not really. The first definition suggests OpenAI aspires to construct tools that will *replace* humans in the workplace. Now, at least as of the time of this writing, the efforts to create mass unemployment through technological innovations have been more hype than substantial threat. But even the ambition itself is ignoble. Were OpenAI to succeed in constructing AGI, the implied end goal would be realized when the corporation finds itself in control of *most* of the economically valuable means of production.

And if we were to judge OpenAI's strategic intentions on the merits of the second definition of AGI, even a generous reader would see in it a very dark agenda. By its own words, OpenAI is committed to the mission of building digital systems that are smarter than organic beings. There are imaginable worlds where artificial creations smarter than humans serve to lift everyone up. But this is not what Altman seeks. Building AGI is described as an end goal, rooted in a sci-fi hope of ultimately bettering humanity. But the dystopian reality of OpenAI's ambition, according to the latter definition, is to work toward ultimately replacing humans not just in the economic sphere but also in the social and political arenas as well.

These are not ungenerous readings. Altman has begun openly opining that his hope for AGI is that it will have roughly the same intelligence as a "median human that you could hire as a coworker."[9] He literally is working for a future where "median" (read: regular) humans are replaced by AI. Writer Maggie Harrison pointedly observes that a "squishy" term like "median" feels like a deliberately dehumanizing choice of word.[10] How can Altman, or anyone else, determine a holistic definition for this statistical average?

It's tough to not see the framing of these AGI-related business ambitions as antihuman. And Altman's subsequent venture, Worldcoin, seems even more dystopian. The new product being offered is a "World ID." Its purpose is to "prove" that the bearer is a real human and not an AI bot. To get the ID, customers need to travel to a dedicated kiosk to *have their eyeball scanned* by a company representative. This is a far cry from "tell me what is in your cupboard so I can tell you how to make a soup."

The early adopters of Worldcoin are being asked to give up their biometrical data, in some markets in exchange for a WLD cryptocurrency token (an illegal offer in North America at the time of this writing), to prove their *humanity*...to a very powerful company...that hopes to eventually render all humans economically obsolete. Altman believes he is softening the shock of Worldcoin's ask by telling the cynical how "people will be supercharged by AI, which will have massive economic implications... We think that we need to start experimenting with things so we can figure out what to do."[11] Marvel at the articulateness of the billionaire class. The messaging is on the lines of "Settle down, plebs. You know, we need to experiment...with things...to figure stuff out." Not exactly a reassuring statement of purpose to justify this type of radically disruptive, future-proofed hustle of global proportions.

Transhumanists and Their Spiritual Machines

The battle between the artful and the algorithmic is ideological in nature. It is a disagreement on how intelligence is to be defined, and what sorts of intelligence humanity should privilege. It is also a disagreement about power, who should wield it and under what conditions. Consider Ray Kurzweil, futurist and director of engineering at Google, stating that the Singularity, the point in time when AI becomes smarter than humans, will happen by 2045. In the aftermath of this event, Kurzweil believes a new breed of posthumans will emerge. Our descendants will be putting AI inside their "brains, connecting them to the Cloud, expanding who we are."[12]

While Kurzweil understands consciousness as the result of biological processes, he nonetheless sees it as a pattern of matter and energy that persists over time and can be transferred to the cloud, allowing us to become what he terms "Spiritual Machines." And he's working hard, with fellow true believers in algorithmic supremacy, at one of the world's most powerful companies, to make this future a reality...or at least convince us that it is inevitable.

I am not in any way positively moved by Kurzweil's transhumanist vision or agenda. But writer Meghan O'Gieblyn offers a personal take on her own journey through Kurzweil's prophecies, providing a sympathetic view into a very different mindset:

> *For the length of time that the Kurzweil book was in my possession, I carried it with me everywhere... It would be no exaggeration to say that I came to grant the book itself, with its strange iridescent cover, a totemic power... .*[13]

> *In hindsight, what appealed to me most was not the promise of superpowers, or even the possibility of immortality. It was the notion that my interior life was somehow real... .*[14]
>
> *To conceive of my selfhood as a pattern suggested that there was, embedded in the meat of my body, some spark that would remain unspoiled even as my body aged—that might even survive death... What makes transhumanism so compelling is that it promises to restore through science the transcendent—and essentially religious—hopes that science itself obliterated.*[15]

O'Gieblyn is unique in explicitly articulating how the possibilities of transhumanism feed into the empty spaces of her soul, gaps created after abandoning religious faith. Even though she has moved on from transhumanism as well, her words testify to the power of this business model. Because make no mistake, that is what it is right now. It is a business strategy, a philosophical worldview whose utopia can be realized only through corporate efforts. We live in an age where science has turned many folks off from traditional religions—and as O'Gieblyn notes, it can be invigorating to be sold the idea that scientific research in a business setting can bring back all the wonders of a spiritual life that were lost.

In reading O'Gieblyn's memoir, I was struck by how different our spiritual journeys had been, even as we both started out in fundamentalist communities of faith. I wonder if the reason I never suffered through a similar crisis of faith can at least partially be attributed to the fact that my desire to remain in a traditional community was only partially rooted in rational calculation. If one day we were to discover that a superior intelligence has been recording the entirety of human history, and in the replay, I saw that the Revelation at Sinai as recorded in the Bible did not happen...it would change nothing for me.

Unlike the surveillance capitalists, I don't privilege certainty. I don't defend the **VICE** of my practice in wholly rationalist or scientific terms.

I raise my family with these traditions *because* they are the traditions of people I love, respect, and want to emulate. In a very pragmatic sense, I like the sort of societies that emerge from living within this tradition. And not only am I comfortable with the uncertainty of a faith-based intelligence, but I am also proud of the resiliency that emerges in communities that are constantly questioning and doubting while trying to live a meaningful life.

What would be far more troubling would be to discover that my beloved grandfather did not live the life I saw him living, or that the joyous religious feasts I am called on to share with my community members are a consequence of us being herded by some malevolent human engineer. The only truth that would cause doubt would be if *my* memories or sense of reality turned out to be false or manipulated. Knowingly living a lie is not the same as welcoming uncertainty. I can live with discovering that my people's collective mythology may be more rooted in ethical aspirations than empirical happenstance. I cannot live thinking my feelings are not my own.

These types of differences came to the cultural fore recently when noted writer Aayan Hirsi Ali announced her conversion to Christianity.[16] Ali was born Muslim but became a very vocal activist in contemporary atheism. Her conversion message was not received well in Muslim, atheist, or some Christian camps. Ali found in Christianity a community within which she can better defend Western values, and practices that can offer spiritual solace in an increasingly cold and empty world. Christian thinkers like Ross Douthat[17] and Andrew Sullivan[18] criticized the lack of apparently authentic Christian experience in her conversion story. Where was the mention of Jesus, the Creed, the Trinity, or the Resurrection in Ali's moment of revelation? Atheist thinkers[19] offered a similarly themed critique, saying they only read in her statement a political stance, not an embrace of religion and refutation of atheism.

I'm outside both of those communities, but to me, these complaints seem overly rationalist and algorithmic. She defends her **VICE**, and I'm here for it. Maybe it doesn't follow the pattern of others, but I still see so much beauty in her statement. At a time of digital essentialism, Ali is drawing on her very human lived experience. If that humanness is now fed by joining with other Christians in meaning-making activities, I don't see why it needs to be subjected to some sort of intellectual purity test. She's shared her intent, choices, and explanations. She's made herself vulnerable and open to critique. That should be enough.

Half-baked Beginnings?

Oxford philosopher Nick Bostrom is currently the most articulate mouthpiece for the transhumanist movement, despite a troubling past of racist pronouncements[20] that should raise disqualifying questions about his moral character. Bostrom defines[21] "transhumanism" as an "interdisciplinary approach to understanding and evaluating the opportunities for enhancing the human condition and the human organism opened up by the advancement of technology. Attention is given to both present technologies, like genetic engineering and information technology, and anticipated future ones, such as molecular nanotechnology and artificial intelligence."

On the surface, this may seem reasonable enough. But Bostrom goes on to note that "transhumanists view human nature as a work-in-progress, a half-baked beginning that we can learn to remold in desirable ways." Coupled with his earlier problematic pronouncements, it is not unfair to wonder if he believes the "work-in-progress" of humanity needs to be differentiated on racial lines. But even in a liberal spin, the transhumanist vision is antithetical to our message, that natural intelligence needs to be celebrated, valued, and centered. If indeed human nature is merely a "half-baked beginning," then we should not be trusting our own minds over the potential of the advanced tools Big Tech is selling.

We should be eagerly seeking an escape from our current state of being, as he, Kurzweil, and other transhumanists and algorithmic supremacists dream of.

Do you see how neatly the five-step hustle fits into the transhumanist operational paradigm? What is a more complex positioning for a product than selling humanity itself as a half-baked beginning to an inevitably superior future? What better way is there to negate the future desired by regular folks who are quite content in their humanness? How can policy-makers push back once they get roped into the vision that this is all about technology helping humanity?

Bostrom opines in a very carefully worded statement that while the transhumanist worldview does not *necessarily* mean privileging the needs of future posthuman beings over current living human beings, it does mean accepting the idea that "the right way of favoring human beings is by enabling us to realize our ideals better and that some of our ideals may well be located outside the space of modes of being that are accessible to us with our current biological constitution." In other words, what is best for human beings today is supporting decisions that will allow us to access a future outside our bodies, outside our minds, outside our existing notions of the self, controlled by Big Tech.

Something akin to a new religion has taken hold among the techno elite. Transhumanism, rationalism, effective altruism, and longtermism are but a few of the doctrines that are now handing down revelations. We already explored transhumanism and rationalism. Effective altruism (EA), a related ideology, is the antithesis of our argument to keep ethics simple and human. In the movement's own words: "Everyone wants to do good, but many ways of doing good are ineffective. The EA community is focused on finding ways of doing good that actually work… It seems that some ways of doing good are over 100 times more effective than others."[22]

As with transhumanism, the polished public-facing mouthpieces for

EA carefully choose words that seem reasonable enough. But a closer look reveals cause for worry. Effective altruists celebrate moral value number crunching as a sacred activity—the more complex the algorithm, the higher its spiritual worth. They seek to maximize the good by only supporting philanthropic causes that can be calculated as offering the largest return on investment, which is what they mean by the seemingly innocuous observation about effectiveness. Of course, only an algorithmic supremacist would be so confident in their ability to accurately predict and capture the myriad of future variables that would be necessary for such a calculation. This type of confidence is best described as a Grand Canyon–size leap of faith.

They also assert the loaded "we can do a lot of good with our careers." Ethically minded people should be choosing professional paths in the financial services sector or Big Tech, because those are the most lucrative. The philosophy suggests that taking a job with an investment bank is a *more ethical* choice than working humbly on the ground to improve conditions for a local community. The logic here is that if you make more money, then you have more money to give away and can subsequently do more good in the world. But the pursuit of money in and of itself, as opposed to the pursuit of an artistic or innovative objective that can lead one to earning money, tends to not work out so well, as folks get caught in the myopia of greed. And the notion that it is somehow *unethical* to choose to take a job at a nonprofit or an arts organization because salaries are lower sets the stage for a very ugly future.

One of the most famous EA adherents, and biggest donors to its cause, is disgraced crypto king Sam Bankman-Fried. He met EA's leading philosopher, Will MacAskill, in 2013 as an undergrad at MIT and was convinced by MacAskill that he could maximize his impact by taking a high-paying finance job, a strategy MacAskill calls "earning to give."[23] On November 3, 2023, Bankman-Fried was convicted of committing an $8 billion fraud.[24]

And while you may not think it possible, longtermists take EA to an even further extreme. Its adherents have concluded that since there will be exponentially more people in the future, particularly in a transhumanist future when next-gen humans will be widely populating digital spaces, improving the lives of tomorrow's more than humans is rated as society's top priority. That means many lives today are seen as less valuable than the lives of those in an imagined future, and the right to life for many vulnerable populations currently suffering can be dismissed out of hand.

On the organization's official message board, Vasco Grilo, organizer of EA Lisbon, muses in an article, "Are we confident that superintelligent artificial intelligence (SAI) disempowering humans would be bad?... From the point of view of the universe, I believe one should strive to align SAI with impartial value, not human value. It is unclear to me how much these differ, but one should beware of surprising and suspicious convergence."[25] Philosopher Émile Torres sums up the impact of all these ideologies with the wry observation that they "have given rise to a normative worldview...built around a deeply impoverished utopianism...who now want to impose this vision on the rest of humanity—and they're succeeding."[26]

Ethics or Alignment?

It needs to be emphasized that while this extreme type of transhumanist thinking remains a fringe ideology in wider society, its adherents are overly represented in the technology industry. For example, Sam Altman is certain that our algorithmic-based operating system is not materially different from that of his digital creations. Consider his response to critics who point out that ChatGPT's output of statistical patterns is nothing like human speech: "Language models just being programmed to try to predict the next word is true, but it's not the dunk some people think it is. Animals, including us, are just programmed to try to survive and reproduce, and yet amazingly complex and beautiful stuff comes from it."[27]

This is an incredibly reductionist view of humanity. To Altman, art, language, and society somehow emerged from the simple algorithmic programming of "survive and reproduce." But "survive and reproduce" does not explain the aesthetic wonders we have brought to this planet, built or created by artists with nobler ambitions than trying to secure a mating opportunity (although certainly some may have been and continue to be motivated solely by the opportunity to reproduce). We all must eat to survive, but that doesn't explain our culinary imagination and the delicacies we have come up with through the generations to bring beauty to food. We have a biological urge to reproduce, but that doesn't explain our achievements in educating the young, creating schools that nurture the soul of the child as well as the mind, filling them with a deep sense of purpose and self-worth. It doesn't explain the choices of parents who not only stick around to see their children grow but also make sacrifices to raise their kids to be better than they are in every notable way. "Survive and reproduce" doesn't explain all the communal structures built to make people feel less alone, our efforts at friendship, increased empathy, expanding the "us" and shrinking the "them."

These techno-ideologues are also working to reframe the question of AI ethics into an "alignment" challenge. "Ethics" looks at principles and is concerned with big-picture explorations about how we create the world we want to live in. "Alignment" is far narrower in scope. AI alignment seeks to ensure that the algorithm advances the original programming goals in the manner that the human programmer intended. That is insufficient, because when innovation is not regulated by the government but left open to market mechanisms, we need to be mindful of the moral character of the programmer.

Bostrom's paper clip thought experiment[28] is a widely cited example of the alignment challenge. A powerful AI programmed to make as many paper clips as possible, using all the resources it can access, at the expense of every other consideration, may start converting all available matter, including human beings and the planet itself, into paper clips. Thus, the AI's execution of its programming will be misaligned from the intentions of the programmer, who would have expected limits on the paper clip production even while not explicitly stating as much.

Focusing political, corporate, philanthropic, and activist energies on this implausible future scenario of existential risk, instead of on the real harm being caused today, is yet another manifestation of the five-step hustle. "AI alignment" is a new field, aiming to assure that future AI systems operate in a way that is aligned with human values. It's a way to position the problems facing tech companies as more complex than old-school ethics.

Alignment is about seeking an efficient solution to the potential challenge of misaligned incentives. Making a show of their efforts to tackle the alignment problem, which is, by the way, ultimately the search for a tech-based solution to a tech-instigated problem, obscures and distracts from how these corporations are currently creating financial value. Worrying about a super AI going rogue and killing every human because it misunderstood its task is more important, they argue, than fixing the AIs of surveillance capitalists that are doing the herding and conditioning exactly as their programmer intended.

When it comes to describing what a business is setting out to do and how it will profit, simpler is always better.[29] Organizations helmed by people who are authentically committed to doing the right thing never have to massage their messaging. The idea that the ethical issues arising from business activities are too complex for the average person to understand was birthed by the crisis management industry for serious gaslighting.

Socially responsible corporations don't feel the need to "start experimenting" on their customers or refer to biometric data as "things" to play around with. Leaders who have the best interests of the population in mind would never hustle the ill-informed into giving up what is most private without seriously well grounded justifications and evidence of high-level protections in place. The major threat imperiling society's future well-being is less in the nature of any AI, flawed as these tools may be, and more in the mindsets of influential algorithmic supremacists. We offer a different mindset.

5

A Brighter Future

LIGHT METAPHORS

While George de Mestral was maneuvering through a forest in the Swiss Alps on a hunting expedition in 1941, his mind was focused on environmental alertness.[1] His hunting companion, an Irish Pointer, yelped in discomfort, diverting de Mestral's attention to the masses of burs from a burdock plant that had adhered to both the dog's fur and his own clothing. He plucked one of the bristly flower heads, now curious about the hooks, which he started to examine through touch between his fingers. De Mestral took this gift of the forest back to his lab to further scrutinize under a microscope. He discovered more details about the hooked properties of these burs, which would stick to anything that had a loop. And while it took years of work to move from this moment of inspiration to a marketable product, artful intelligence led to the creation of Velcro. De Mestral's experience of thinking with body, environment, action, and mind—**BEAM**—helped spark the invention.

The metaphorical implications of the **BEAM** acronym are obvious. Light permeates our language—we beam in response to good news, experience moments of illumination, glow with enthusiasm, show flashes of brilliance, offer rays of hope, etc.[2] Innumerable cultures assign the highest good to beams of light. In the Judeo-Christian spiritual tradition, the first act of creation emerged from God uttering "Let there be light." As Rabbi Adin Steinsaltz explained, "Light as a positive symbol is so prevalent in biblical Hebrew that redemption, truth, justice, peace, and even life itself 'shine,' and their revelation is expressed in terms of the revelation of light."[3]

Learn to use the entirety of your **body** to satisfy a sense of curiosity. Toss in the wondrous array of resources that are at your fingertips, generally supplied by the unique physical **environment** you find yourself in. The hope to outsmart AI comes from realizing that our cognitive processes get most fired up when we take goal-directed **action,** or to put it more crudely, when we get off our asses and actually get shit done. If you want to think better, do something impactful in the real world; stop viewing your brain as a computer—expand the boundaries of what you call your **mind**. Big Tech is working overtime to devalue our inner light. The hopeful response is focusing our **BEAM** outward and radiating brightly. If these metaphors are a little bit out of your comfort zone, buckle in. Just wait until Part III, when we explore the power of love. But to keep your attention until then, here's a confession: I continue to struggle with fully internalizing this language, these concepts, and immersing myself comfortably in an artful reality.

Unlike the folks who we will be meeting in Part III who naturally tap into their bodily wisdom, I've found it to be a long journey arriving at the artful worldview. **BEAM** is the best reaction to Industry 4.0. But that doesn't mean it's an easy intellectual shift. Most of us have been subjected to decades of programming designed to turn us against our bodies and minds. It was not an accident that developing societies were trained out of the varied methods of problem-solving successfully employed throughout preindustrial history. Devaluing natural human judgment, like engaging in trial and error or trusting our instincts, and selling strict rule-based approaches as the only reliable way to think served the agenda of those who stood to profit most from the shift—people in the business of control.

My Brain Doesn't Trust My Body

The message of mind/body dualism, the worldview of rationalists who hold the mind as primary and bodies as untrustworthy casings of secondary value, was everywhere in both my religious and secular educations. Reflecting on those formative years, I still feel in my gut the resonating security brought by dualism. And, in keeping with this book's philosophy, I'm trusting that messaging. It's why I'm putting everything on the proverbial table. There is much good in rationalism and a scientific worldview. But as with many good things, moderation is key.

In early life I existed as a nerdy, skinny kid who felt safest and happiest retreating into the mind. My preferred pastimes were reading, especially religious texts; studying; or listening to music. I hated sports and other athletic activities that put the body front and center. What I was taught by both rabbis and professors made perfectly rational sense. It felt good, and true, to conceptualize the mind as housing "the real me," while the body I was born into was an apparatus of secondary importance. My physical reality was necessary for navigating this temporal plane but ultimately a hindrance to the "noble" scholarly work I wanted to do. The "real me" found joy and meaning in refining the intellect and what some of us (still) call the "soul." Others call it the essential "higher self," while Kurzweil calls it a "pattern." What's in a name, right? Whatever you call it, it's the piece of being, perhaps metaphysical but more likely biological, currently trapped in but still somehow detached from the body.

As a youngster educated in a Yeshiva day school, I eagerly absorbed teachings that positioned religious practices as rites designed to take "debased" physical activities and turn them intellectual. For example, making a blessing before eating a sandwich was a way to turn the mundane act of nourishing the body into a spiritual exercise of intellectual mindfulness. I loved this explanation and relished the opportunity to feel good by making everything I did an act of the mind. We learned Talmudic discourses exploring why it was forbidden to derive any benefit from the lowly physical world without consecrating the act or item by reciting a blessing.[4] Satisfying bodily needs was never presented as intrinsically bad—Judaism is not an ascetic religion—but sidestepping the opportunity to intellectualize the experience was a selfish choice. We could make the physical holy, and refine our souls, simply by using the brain as an intermediary.

And while that is where my journey started, it is not where I am now. Mercifully, just as the rabbis programmed me to quiet the wisdom of my body, it was also a rabbi who snapped me out of this lull. I've told a version of this exchange previously,[5] but it's too critical to what we are discussing here to not revisit.

During my undergraduate years, I was blessed to have the mentorship of Dr. Yitzchok Block, a rabbi and philosophy professor. Dr. Block was a brilliant academic, earning his PhD at Harvard and writing a book during our time together reconciling rationalism with religious faith. He was also a devoted Hasid of the Lubavitcher Rebbe who would dance through the streets of London, Ontario, and, more peculiarly, dance on the tables of the university's Chabad house, engulfed by mystical rapture. The rational side of him I understood. But this penchant for table dancing confused me and made me extremely uncomfortable.

We spent years arguing about the content of a life well lived. And, the truth is, I loved all sides of him—the expert philosopher who would captivate the lecture hall with his insights on moral philosophy, and the wild-eyed crazy person my roommates would let into my bedroom in the early hours of the morning to drag me, quite literally, out of bed when Dr. Block needed a tenth person to form a quorum for morning prayers. But I only *understood* the intellectual side of him; it was the scholarly persona that I found aspirational.

Our back-and-forth on identity and purpose continued until one day, we were sitting at his dining room table, having consumed copious amounts of vodka, and he stared straight into my eyes and said, "You will never know the incredible joy I feel in being a Hasid of my rebbe. Your overdeveloped intellect will never allow you to feel what I feel." And with that he poured us another shot, and we never debated the issue again.

Because…he was right. What could I say? There was no reasonable response to this deep insight. His rebbe, Rabbi Menachem Mendel Schneerson, had once described the rabbi's role as a geologist of the soul.[6] There are many treasurers in the earth, he explained, but if you don't know where to dig, you'll only find dirt and rocks and mud. The rebbe can tell you where to dig, and what to dig for, but the digging must be done alone. In that moment, Dr. Block was a geologist of the soul. He was telling me where I needed to dig to uncover the treasures of a life well lived. But to get that life, there was a lot of inner work I would need to do first. I lived in my brain. Overanalyzing and hyper-rationalizing kept me from seeing the beauty that could be accessed only by embodied wisdom.

Not to disappoint, but I still don't dance in the streets. Or on tables. Or anywhere, to be honest. And it's to my detriment. Recent research has discovered that dancing is a more effective treatment strategy for depression than prescription medicine, exercise, meditation, or any combination of those activities.[7] Dancing. I'm clearly missing out. I still don't know the deep joy of religious ecstasy. I still don't trust my body. But I've been digging for many years now, doing the work. Because bodily distrust has had a demonstrably negative impact on my ability to think, heal, and thrive.

Body over Mind

For some reason, the message that often gets shared is "mind over matter," meaning our thinking can impact physical reality. This is true, but so is "body over mind." We can change the way we think by changing the way we are sitting or changing our patterns of breath. Sufferers of anxiety disorders know how bodily shifts can redirect a mind locked in fight-or-flight mode. It's often a thought that triggers a panic attack, and the best responses to these cognitive misadventures are physical: grounded feet, supported back, tongue to the roof of the mouth, head up, eyes to the horizon, etc.

In fact, so much of what works in therapy is body based, not mind focused. It's not the ability to have a witty or open verbal back-and-forth with a therapist that indicates success. It's the extent to which a client and their therapist mirror each other's body language that signals the strength of their bond.[8] This mirroring might include coordinated body movements and gestures, the congruence of their posture, mimicry of each other's facial expressions, or the adoption of similar voice quality. It's not the words that matter most. It's the embodied connection. Is this surprising? Have I ruined therapy for some of you, who will now be overly self-conscious about how you present at your next appointment?

Our bodies give away all our secrets. Do you want to know if you can trust someone you just met? Introduce yourself with a handshake and pay attention to what you feel in your palm. Psychologists exploring the science behind the "warm hands, warm heart"[9] aphorism found connections between physical warmth and the psychological state of being warm, that is, friendly and trusting to others.

Further studies discovered that people who have experienced being socially rejected rate the temperature of the room they are in as colder than do people who haven't.[10] Think about that the next time you are reaching for a sweater. Take a moment to be curious about whether the best resolution to your feeling of coldness may be social.

Want to get a good read on whether you've pissed off a friend? Before you next hang out together, down a magical elixir of spicy soup. Eating spicy food affects our emotion perception process, allowing those who consume heat to be more aware of facial expressions signaling anger and disgust.[11] Similarly, if you want a romantic outing to go better, head to the dessert spot. Experiencing the taste of sweetness has been linked to prosocial behavior and the perception of people as more attractive.[12]

For many of us, our moments of greatest emotional vulnerability take place with food, around a table. There's good reason to be curious about how what is on the menu may influence our bodily experience. I am not sure that the programmers of the soup-making GPT from the start of our book factored this reality into their algorithms, although I am certain that the human engineers of surveillance capitalism do, which makes it even more important that we do as well. It's very tough to encounter all this evidence and not start to recognize the major role that our bodies and physical surroundings play in affecting, supporting, and enhancing how we think.

I share this in the hopes that the more rationally inclined reader will stick with us a little longer. There may be a temptation to jump past these prescriptions. I get it. The allure of modern rationalism is in the tale that the strong-minded can overcome the weaknesses of their bodies. That progress is infinite, as we will always be able to innovate our way out of a problem. Our moral choices reflect our success at tapping into an innate truth, rather than learning what is right from the people we respect. Big Tech has been good at navigating this revolution. But try to stay open to the possibilities built deep into the prescriptions you'll encounter here, despite the perhaps displeasing aesthetics in their positioning.

It is nothing less than bad science to describe human cognition as a process taking place exclusively in the brain. AI built on a neural net model, where data is processed in hardware that uses interconnected neurons in a layered structure, is claimed to possess architecture that resembles that of the human brain. But that's not how the human brain works, nor is the brain the sole site of thinking. Those of us whose futures aren't tied to selling superintelligent machines ought to start paying closer attention to scientists describing human cognition as a multiscale web of information processing running across the entire body.[13]

Computer engineers want to "build a brain" so that they can understand their own brains. But while they have created something wondrous with AI, it in no way replicates human intelligence. Neuroscientists admit we have yet to figure out how brain activity generates cognitive and mental states. A helpful clue, however, emerges from a consideration of how neurons, a type of cell, work in tandem with other types of cells in the body, like those of the immune system. These experts are more interested in holistic, full-body interactions than the limited focus on activity in only one organ (the brain).

The brain is part of the body—an observation that gets lost on those caught up in dualism. Despite the wishes of many, and I was once one of them, our brains are bodily organs, not a piece of the self that can be meaningfully detached or differentiated from the rest of us in a fundamental way. And like all other bodily parts, the brain is made of cells. So why aren't we looking more closely at cellular processes and interactions? While neuroscientists are reconsidering the unwarranted belief that the brain in isolation is the home of mental states, Big Tech's AI innovators are blindly sticking to old science.

As for us regular folks, we can learn to trust something about ourselves other than our calculating rational mind. I know from firsthand experience how difficult it is for those trained to trust our brains over all else to make this leap. So start small. Like, trust your breath. Breathe in deeply, hold it, exhale slowly, and accept it as sufficient testimony to know that you are alive, that you are present, that you have opportunity. For me, this is not a trite prescription. There are still far too many moments when I do not trust my breath. It's a process. It takes time. But it's movement in the right direction.

And if in this suggestion you hear echoes of mindfulness or prescriptions from pop psychology and think "that's not for me!" well, you've proven the point. Get out of your head. Quit with the labels. Get reacquainted with your body. Start slowly trusting your bodily instincts when your brain tells you to calculate and follow the algorithm in a different direction. Get into the messiness of trial and error to solve a problem, even if you have a rule-based approach. Go so far as to break an algorithmic rule when it feels right. This is the start to growing your capacity for **BEAM** thinking.

Brainy Bodies

Algorithmic supremacists are working to build a future more amenable to the needs of posthumans rather than make life better for those living today. Artful intelligence is constructed on cutting-edge research,[14] which is discovering the myriad of ways intelligence manifests through the cognitive processes that take place in the experience of having a physical **body**.

Earlier in the book, we adopted the APA's definition of intelligence as an uncontroversial starting point: the ability to derive information, learn from experience, adapt to the environment, understand, and correctly utilize thought and reason. Yet revisiting it now, after learning more about embodied cognition, we can see how burdened that definition is by the weight of dualism. Look at the back end: "correctly utilize thought and reason"—correct by what standards? And why privilege reason over other cognitive capacities?

At this point, we may be better served by a more holistic definition. Moving forward, let's work with cognition defined as "the sensory and other information-processing mechanisms an organism has for becoming familiar with, valuing, and interacting productively with features of its environment in order to meet existential needs."[15] Notice the move from the curt analytics of "derive information" to the more embodied "sensory and other information-processing mechanisms."

Developmental psychologist Esther Thelen adds that what we call cognition not only involves bodily processes we are not conscious of, but it also arises from our bodily interactions with the world.[16] Our very ability to think is contingent on experiences that can come only from having a physical body, and that unfold in specific ways because of our distinct perceptual and motor capabilities. These abilities, from the capacity to reach out and grasp an object that's in front of us, to the capability of hearing the sound of thunder from a flash of lightning that is many miles away, are inseparably linked. Together they form what Thelen describes as "the matrix within which memory, emotion, language, and all other aspects of life are meshed."

Most of today's tech leaders embrace an absolute distinction between body and mind. Geoffrey Hinton, widely named as the "godfather of AI," recently gave a talk[17] at Cambridge University where he mused, "I don't think there is anything special about people, other than to other people." Hinton rejects the notion that an intelligent being needs to "act on" the world physically to understand it. Whereas in the new definition of cognition we have adopted, acting is essential. Becoming familiar with, valuing, and interacting productively with features of one's environment to meet existential needs is the critical feature of intelligence as we define it.

Hinton also has little interest in consciousness, a concept he views as "pre-scientific," noting that in his opinion "understanding isn't some kind of magic internal essence. It's an updating of what it knows."[18] Once again, by reducing human intellect to computational features like a software update, Hinton can advance the self-serving notion that a chatbot has the same capacity for understanding as humans do. "There is no mental stuff as opposed to physical stuff," he says; there are "only nerve fibers coming in." From his perspective, all we do is react to sensory input.

When given a platform for their contemplations, the great minds behind AI innovations (I say this seriously—they are great minds even as they are fallible) who push for algorithmic supremacy reveal an absolute blind spot in their curiosity when it comes to being embodied. Here's Hinton on a completely different topic displaying the same biased mindset: "I often think about it when I'm not sleeping at night. There's something funny about sleep, which is some animals do it, fruit flies sleep. And it may just be to stop them flying around in the dark. But if you deprive people of sleep, then they go really weird… What is the computational function of sleep?… I now think it's quite likely that the function of sleep is to do unlearning on negative examples."[19] Hinton is only curious about the "computational function" of sleep. He homes in on the singular salient variable to one who views the brain as a computer. What about the possibilities that sleep recalibrates our emotions, restocks our immune systems, fine-tunes our metabolisms, or regulates our appetites? All those silly things that go along with having a body.

Further, Hinton's conclusion assumes that the sleep state serves the wake state. After all, the brain is only interesting to these folks when it is "on" and actively computing. But researchers in neuroscience theorize[20] that it is possible that the functions of rapid eye movement (REM) and non-rapid eye movement (NREM) sleep have more to do with the sleep states themselves than with waking consciousness. REM may be undoing something that NREM is doing. Or conversely, NREM may be doing something important for the body, like immune system repair, but that function is costly, so REM functions to complete, complement, repair, or undo something that NREM had to do to accomplish its primary functions.

The point in sharing all these Hinton quotes is not to simply pile on him for controversial views. It's to demonstrate clearly, with a living example, what we stated from the beginning—that AI is an ideology. Hinton was a powerful player at Google and since he retired has been a go-to source for media interviews on AI. His influence on the development of certain products won him a Nobel prize, and for that he should be lauded. But his ongoing influence in shaping the ideology calls for pushback. More artful curiosity and a little less hubris, a little less of a commitment to certainty, are what's warranted.

Let's center alternate views, like that of author and cognitive scientist Guy Claxton, who remarks, "what goes on in our bodies...are the roots and trunk of all the other forms of intelligence. Bodily intelligence gave birth to them, holds them firm, and continues to nurture and support them throughout life... If we took away the bits of our brain that are coordinators of bodily processes, there would be nothing left—no brain, and also no wordplay, no poetry, no algebra, and no imagination."[21] With these new parameters of meaning making now set out, what, then, is the first ask of artful intelligence? What does it mean to adjust, in a practical way, our lived behavior to accommodate and integrate the knowledge that thinking is not limited to the confines of the brain, but takes place throughout the body? What should we start doing differently, right now, in this moment?

When you next find yourself most in your head, repeat this mantra: *I'm aware of my body telling me what to do.* To think with **BEAM** is to be mindful of the fact that we are always acting on bodily senses, in any given situation, no matter how badly we may want to frame it as otherwise.[22] That blessing of being human should be amplified, not repressed. AI boosters envision a future where every human act is guided by information provided by a technological enhancement, not the unique, in-the-moment sensations of an embodied being who is thinking, in a holistic way, about what to do next.

Big Tech's engineers assume that our best decisions are products of rational algorithmic calculations processed by the brain. They believe that the only way for humans to make better decisions, and to help usher in a posthuman future, is to augment us with sensors allowing access to data that will be provided by superior algorithmic processors. Machines will allow us to overcome our biological limitations, enhancing the human capacity to think, and leading to a world filled with better outcomes for everybody. This is the religious belief expressed by O'Gieblyn, that we can retain the "spark" embedded in the meat of our bodies while outfitting it to become a spiritual machine.

But many of us are still happy being spiritual humans. To proudly think with the body (because that is what is happening whether we choose to be mindful of it or not) is to push back against the marketing campaigns of transhumanists. Their products are designed to promote efficiency. Efficiency in the economic sense is an assessment of the relative value of the ends achieved given the costs of the resources consumed to get there. Despite the technical language, measures of economic efficiency are notoriously subjective. Who decides the value of the inputs and outputs?

At the time of writing, Apple, Meta, and Google were in a race to finally create a headset with mass-market appeal. They want to sell you on a future where, for instance, you will be wearing smart glasses on a date, because a chatbot informing you of what to say next is so obviously a more efficient way to interact with another human. They are priming you to believe that you are operating on robot time, where there is no allowance for error or setback—where getting the job done, establishing a connection, is mission critical. But I'm sure that's not what your heart tells you. The headsets are designed to isolate your focus and keep your attention away from your heart and gut.

This messaging is especially resonant if the human engineers have been successful in lowering your defenses and sense of self-worth, convincing you that an average and unimpressive person such as yourself (not *you*, though, dear reader) is bound to disappoint in these high-stakes social exchanges. From the perspective of human engineers, it would be optimal for both parties on the date to be equally tech-equipped, allowing for the superior insights of an AI to efficiently navigate the entirety of the human courtship process. Exploiting the natural insecurities most of us have in intimate social interactions and selling us tools that promise to make these exchanges more efficient ultimately removes the humanness from these experiences. And, not by accident, it increases the power and control of those selling these tools as we eventually forget how to court, converse, and relate.

BEAM thinking will help in retaining these human qualities at a time when everyone around you is under the algorithm. The practice of being more artful in social settings can start with the simplest of exercises, like paying closer attention to the subtleties and nuances when engaged in less stressful conversations. Work to become aware of how you are naturally guided to keep the back-and-forth moving by a bodily sense of the best way to proceed. Pay attention to how you, and the person you are interacting with, are using eye contact and touch, gestures and altering tones. See how, despite insecurities, you can successfully navigate to the other side of moments when maybe you didn't love the words that came out of your mouth but all was not lost.

Become more mindful of the many different roles and functions your bodily actions can play. For example, gesturing in conversations is not strictly communicative. In fact, these movements often serve a cognitive function for the speaker, helping to smooth out their thought process and assist in getting the words out.[23] Knowing this should be very empowering. For while we may not have the clean operational efficiency of an algorithm using probability to predict what words should come next, we have a wondrous effectiveness in bodies that help us get to where we need to be.

Go with the feeling that something is missing or not right in an interaction. Don't retreat to solitude or a screen, but use your hands, your face, your heart, and any other bodily resource to try to figure out what needs to happen next. Pay closer attention to those feelings, the messaging they mean to project, and trust the bodily sensations that will let you know when you eventually land on the right response that supports forward momentum.[24] We can learn a lot in those moments and become better at, well, life.

I'll say again that I'm very much aware that many of these prescriptions are easier said than done. They involve a substantial amount of self-awareness, and we've been conditioned to keep our eyes outward to the screens, not inward. Maybe to help get us closer to this state, a type of somatic investigation would be helpful. If you're not ready to check in with your feelings, try checking in with your body.

As you read these words, are your shoulders rising? Is your jaw clenching? Work on this first. Look out for signs of physical discomfort and see if you can ground yourself enough to comfortably work through it. Can you get your shoulders down? Can you relax your jaw? Can you quiet your mind if it is telling you that artful thinking is for the hippies and artists, not the serious people like yourself? If it's any comfort, I'm still not fully comfortable with this. But the work is worthwhile. The more fully you experience your body, the more effective you will be with artful thinking.

Use What's Here

The next step in **BEAM** involves a more robust awareness of the **environment** around us. Access all the tools and resources afforded by the material spaces you find yourself in at any given moment, like the hunter/entrepreneur at the start of this chapter who paid close enough attention to get curious about a plant that insisted on literally sticking with him.

Are you one of the half billion people who have played Tetris? If so, seeing the name probably triggers the glitchy soundtrack of a Russian folk song playing in your head as you flash back to the storm of falling blocks needing to be rotated to fit into the spaces left open by the previously fallen blocks. Players need to decide how to orient each block, as well as where to drop it, before the block falls too far. What's interesting for our purposes is how most of us used actual rotation and translation movements to simplify the game's core challenge.[25] I'm sure you have vivid memories of doing the same—rotating the blocks on the screen as they fell rather than picturing the rotation in your head. That was artful thinking—using the physical game space to help work out a solution—as opposed to a computational approach of figuring out the optimal positioning mentally and then executing the moves immediately.

Or, returning to the scenario we started the book with, think of the different approaches to making a soup. When following a recipe, whether generated by an AI or a human, you'd take out the ingredients in very specific measurements, like six cups of broth, one pound of chicken, seven cloves of garlic, etc. But if you want to use the exercise to refine your artful intelligence, try laying out a recipe's ingredients in front of you and apportion a quantity into the mixture based on what you see, not what you were told. Use the physical environment to help you think about a strategy for real-time soup creation.

This process is called "off-loading," where we move some of the cognitive work out of our heads and onto the temporary spaces made available to us in the moment.[26] Cognitive off-loading is using physical action to alter the information-processing requirements of a task, thus making it easier to think about. This simple hack improves performance in perception, memory, arithmetic, counting, and spatial reasoning.[27] The trick is using tangible spaces to hold and manipulate information for us, harvesting the secured data only when we need it to complete a specific task.

So, if it is our intent to think more artfully, we can start to consciously seek out opportunities for off-loading. Maybe deliberately count on your fingers the next time you are faced with simple addition that you would usually do in your head. Or bring out pencil and paper for doing more complicated math. Not because you *need* to, but to train your body into knowing that it is good and natural to off-load.

At this point a critical reader might ask, what's the practical difference between always grabbing a pencil and augmenting our bodies with tools like Apple Vision Pro? Neuroscientist Francesco Ianì warns that every time we take an external object as a tool, processing it as if it is now an extension of our bodies, there is a price to pay. "The real body goes through alterations of different kinds...these changes are all interpreted as costs since, in general, they result in a reduced ability to process both sensory and motor information."[28]

And what does this mean for off-loading cognitive tasks to AI? Researchers find two costs that regularly manifest in safety-critical situations, like flying a plane or driving a car. The first is automation complacency, where we cease to be sufficiently vigilant to the performance of the off-loaded automated processes. The second is automation bias, which is the tendency to uncritically rely on the output of an automated decision aid. Long-term reliance on automated processes, therefore, could lead to cognitive skill decay, where embodied abilities we once had deteriorate.[29]

Like many of you, I see how my navigational abilities have begun to wane since I allowed the Waze app into my life. I justify the decision in the name of efficiency—Waze always has a shorter route. I'm shaving off minutes from every trip. But what price am I paying? Numerous studies have found that regular use of GPS-based navigation apps negatively affects spatial memory when we later try to navigate on our own.[30] Is it worth it?

Thinking artfully means using the environment to support a deep grounding *in* our bodies. To start tapping into this ability, begin making a concerted effort to pay attention to, and be curious about, your immediate surroundings. This tactic is especially important when you find yourself deeply consumed by a cognitive challenge, and your brain is telling you that looking at the physical space you are in, would, in fact, be a "distraction." Resist that algorithm.

Whenever possible, get away from your desk, move your body through physical space, and experiment with how your setting might support you in your efforts to work through the challenge you are facing. Timothy Leary was on to something when he spoke of the importance of set and setting. See if you can find inspiration, or an opportunity to off-load some of the work. Try to make your problem-solving exercise tactile—sort, touch, twist, and flip your way through to a solution.

Thinkers Are Doers

AI innovators like Hinton hold tight to an ideology espousing that the human mind is literally, not metaphorically, like a computer. This is why their ilk have come to believe that the only way to understand it is by replicating it. Their technological breakthroughs were guided by the principles of algorithmic supremacy, confident that research advancements allowed them to crack the code behind human cognitive processing, so they would end up building a brain that their firms could patent and own.

But when we think with **BEAM,** we disrupt that assumed simplicity. We think with a combination of **body**, **environment**, and **action**, something their machines were not programmed to do. And sure, there may come a time when AI has fully integrated sensors that allow its computational process to be a mix of inputs. But even then, sensors aren't eyes, and a processor attached to mechanical limbs is not a body. To fully replicate human cognition, the AI would need a mirror of our physiology, from nervous system to digestion.

A third step to becoming more artful is to stop seeing the **actions** we take as distinct from the process of thinking. Having goals and getting things done starts in the head—we need to know that the moves we are about to make will have the effects we hope for. We enact these imagined processes precisely because we want to achieve that specific predicted effect in the world.[31] Thinking and doing are two sides of the same coin. Ignore any therapist or life coach who tries to tell you otherwise. Goal-directed action is a deliberate and intentional expression of our ability to bring about desired consequences through our embodied mental state.

In laying out the prescriptions for artful intelligence, we have repeatedly drawn on the instruction to be mindful. But this is not the mindfulness of the meditation room. It is mindfulness in action. A popular colloquialism implies that there are two opposing categories of people: "doers" and "thinkers." And in the business world, the former is desired far more than the latter. Unsurprisingly, there are algorithmic supremacists hustling a product that helps separate doers and thinkers more efficiently, so that employers would have a clear metric for how many strategic planners they need on their team relative to the number of action-oriented risk takers.[32] But when you think artfully, this dichotomy makes no sense. Thinking can never be divorced from acting.

Algorithmic intelligence views cognition as being isolated in the computational power of the internal hardware, manifesting as a brain in humans and a processor in machines. Artful intelligence rejects the framing of cognition as computation. The goal of our existence is not, as Hinton believes, simply to process data. We are not meant to sit idly and analyze. We have an internal impetus to respond to our ever-changing reality through actions that unfold in real time. Let's revisit our working definition of cognition: to think is to interact productively with features of our environment to meet existential needs.

Go back to artful soup-making: We use our sensory mechanisms to interact productively with the ingredients in front of us so we can meet our biological and aesthetic needs for a nourishing and delicious soup. We aren't simply following an algorithm to attain nourishment. We are smelling the ingredients, tasting the soup, seeing the changes, and adjusting the concoction according to our specific tastes and bodily inspired whims, because we need to eat but we also need to feel good, create, and be proud of our efforts.

An inevitable outcome of beginning to think more artfully is the reshaping of our understanding of the five senses. In the old paradigm of cognition, the consensus on the purpose of vision was that it helped the brain by building the data for an internal representation of the perceived world. In other words, our sense of sight was used to gather the data of "what" was out there and "where" physical resources were situated. The visual system provides raw information to help solve queries like "Where did I put the soup ladle?" The visual system was simply for mapping and identifying, similar to how optical sensors on a computer are employed. It was the brain, or internal processor, that decoded and made use of this information.

Research in embodied cognition finds that the visual system is better understood as a "how" pathway, designed to support visually guided actions such as reaching and grasping.[33] It's not "Where did I put the soup ladle?" but "How can I use my hands and arms to grab the ladle and taste the soup?" What findings support this? When people were asked to quickly identify whether a frying pan was upright or inverted, the response times were fastest when the response hand was the same as the hand that would be used to grasp the frying pan, like the left hand if the pan's handle was on the left.

Similarly, understanding the function of memory as "for memorizing" is replaced in the artful worldview with the more action-oriented process of encoding patterns of possible physical interaction with a three-dimensional world.[34] In this perspective, we store memories of objects and situations in terms of their functional relevance to us, not "as they really are." This means that short-term memory is better conceived as the deployment of particular action skills.

Say you were heading to the grocery to pick up cucumbers,

tomatoes, and onions. As you drive to the store, you repeat that list over and over in your head. As you wander the aisles, you continue to repeat the list. In fact, "cucumbers, tomatoes, and onions" became your mantra from the second you headed out the door until all three items were placed in your shopping cart. Your working memory didn't activate a unique type of storage for that list; it facilitated the action of verbal repetition, which in turn facilitated the action of acquiring the requested produce. This was thinking in action—repeating new information out loud facilitated learning.

Philosopher Alva Noë looks at how dancing can be conceived as a way of thinking. He explains that dancing is essentially at odds with actions that have been traditionally associated with cognition, like planning and deliberation. When a person is seen as dancing well (we're not talking about a professional dancer here), it is because that person has let themselves go. They have released themselves to the instinctual responses their body wants to manifest in reaction to the rhythms and sounds. They just dance with abandon, without a plan or strategy for movement.

But, Noë notes, dancing is not mindless: "to dance well is to manifest a distinct kind of intelligence; dancing isn't just moving; it is, usually, moving in coordination with others as well as music; it requires attention (to beat, rhythm, phrasing) but also, more generally, a sensitivity to gesture, expression, timing, feeling, social relationship, humor, and more."[35] Yes, you are giving in to your body. Yes, you are abandoning your ego. But you are not abandoning thought. Your actions are clear expressions of a cognitive process. You are engaging the sensory and other information-processing mechanisms of your body to become familiar with, value, and interact productively with the musical and social features of your environment.

In a similar spirit, philosophers Elisabeth Pacherie and Myrto Mylopoulos explain how viewing cognition as undertaken through goal-directed action sheds light on how we acquire a skill.[36] The general understanding is that skills are developed through practice. But that's also the process for forming good habits, and Pacherie and Mylopoulos want to understand the difference between skills and habits. The key lies in the ability to *willfully* link cognition to doing things and achieving our goals.

A skilled move is supported by (at least) three cognitive activities. The first is a mental representation of what it is we want to do; we have some sort of conception in our head of the behavior we hope to enact. This conception is less a photo-realistic snapshot of the skill-in-action and more a notion of what that skill means on a very personal and individual level. The second is a mental representation of action implementation, which is an idea of *how* you are going to do this thing that you are so good at. And the third is an internal model for action selection, a unique cognitive tool that you created to help you, and only you, decide *what* you are going to do.

If you have a skill, it is likely that you have well-structured action representations in your toolbox to allow your body to select better movements, better anticipate action effects and future events, and better monitor and control your performance. Further, your skill likely emerged from very detailed and accurate models of a specific domain at both the strategic and the situational or tactical levels. Having these models allows you to make optimal decisions, even in unusual and challenging situations.

Now, a point of clarification: Incorporating new ideas like associating cognition with doing real things into our artful toolkit does not mean we now have a comprehensive understanding of the enduring mysteries of natural intelligence. Psychologist Margaret Wilson considers examples of visual events like sunsets, which are always perceived at a distance and do not offer any opportunity for physical interaction. We can't reach out and grasp the sun. Or human faces, which most cultures assign as a no-go zone to uninvited touching by strangers. Or the act of reading, where visual pattern recognition is paramount and opportunities for physical interaction with those patterns are basically nonexistent.[37] In sum, arguing that visual systems should be understood as serving the "how" doesn't mean they don't *also* serve the "what." We simply see sunsets, faces, and words. We don't see them to grab and manipulate.

But Wilson explains that our cognitive processes are meant to promote action via a flexible and sophisticated strategy in which information about the nature of the external world is stored for future use without strong commitments on what that future use might be. She explains how walking into a room, she may notice a piano as providing a bench to sit on. But we also have the ability to draw on this visual knowledge in a variety of unforeseen circumstances, like needing a tool to make a loud noise, or a barricade against intruders, or fuel for fire. "These novel uses can be derived from a stored representation of the piano. They need not be triggered by direct observation of the piano and its affordances while one is entertaining a new action-based goal." Artful thinking comes from this amazing ability to use the full resources of our bodies and environments to help us do things we never thought we might have to, as we respond to emerging challenges in real time.

So become more conscious of these sometimes-unconscious processes. Start willfully incorporating pauses into your day for a moment and check in to see the actions you may be taking. For instance, going back to the shopping errand example, be mindful of *how* you are working to remember the list throughout the errand, not just the outcome of returning home with the groceries because you remembered the full list. How do you think through doing? What habits can you change to think by doing more often?

The Mind as Style Consultant

To think artfully is, step one, to think with our **body**; step two, to enhance this process by accessing resources made available to us in the physical **environment** we find ourselves in; step three, framed not as passive computation but to engage in goal-directed **action**. All of this, when considered together, leads us to our fourth prescriptive step: to adopt a holistic and relational conception of the **mind**.

Unlike those who perceive the world as a function of algorithms, artful intelligence rejects the view that our ability to create, navigate, and innovate comes from the power of an isolated brain, engaging in analytic computation, that can be replicated and even upgraded as a synthetic piece of processing hardware. Instead, to think with **BEAM** is to be mindful and proud of the interplay between the mind, an equally important body, a constantly changing external environment, and ever-shifting goals.

As Claxton puts it: "My mind was not parachuted in to save and supervise some otherwise helpless concoction of dumb meat. No, it's just the other way round: my intelligent flesh has evolved, as part of its intelligence, strategies, and capacities that I think of as my 'mind.' I am smart precisely because I am a body."[38]

Noë offers a metaphor that resonated with me: "Thought and perception differ as styles differ. A style is a way of doing something—dressing, writing, singing, painting, dancing. Thought and experience are different styles of exploring and achieving, or trying to achieve, access to the world."[39] Let the mind be understood as our style consultant, empowering each and every human to **BEAM** in their own ways. Style is the essence of human activity, as it describes what we are for all to see.[40] No human has no style, and the artistic attitude is the most natural of expressions.

There is still so much that we are learning about what constitutes the mind, and how our brain and body interact. Neuroscientists had a breakthrough in finding a previously unknown system within the brain located in an area responsible for the movement of specific body parts and engaged when many different body movements are performed together. They named this system the somato-cognitive action network

(SCAN) and showed its connections to brain regions that help set goals and plan actions. The SCAN supports mind/body integration, allowing the brain to anticipate upcoming changes in physiological demands based on planned actions.[41]

These researchers believe that the SCAN enables pre-action anticipatory changes in one's posture, breathing, and cardiovascular system, like shoulder tension, increased heart rate, and "butterflies in the stomach." This can explain why we experience physical responses like sweating or increased heart rate when just thinking about a difficult task we have yet to undertake. Action and body control are melded in a common circuit, providing a neuroanatomical explanation for "why 'the body' and 'the mind' aren't separate or separable."[42]

When Sam Altman, who dropped out of Stanford's computer science program after his freshman year, was recently asked by a fanboy what he has learned about *humans* after working on AI for nearly two decades, he answered: "I grew up implicitly thinking that intelligence was this, like, really special human thing and kind of somewhat magical. And I now think that it's sort of a fundamental property of matter."[43]

Nobody who is serious has considered intelligence to be the sole property of human beings, not shared by other *living* creatures, for quite some time. But more significantly, this "learning" is false. Intelligence is not a fundamental property of matter. Properties of matter are characteristics like mass, volume, and temperature that can be observed and measured in all physical things. There is no trait of intelligence without the ability to learn, and it's contrary to observable evidence to suggest that anything that physically exists can also think and learn. It would not be hyperbolic to state that human intelligence *is* somewhat magical.

Algorithmic supremacist tech leaders like Altman have repeatedly shown a disdain for the inefficient ways humans read, write, draw, compose, and create, wanting us to outsource those activities to machines that can spit out mathematical variations of what uncredited human artists produced with heart and soul. The algorithmic supremacists of Big Tech create the tools to do this and name a value-destroying process nobody asked for as efficiency, progress, and improved productivity. Twenty years in the tech industry have taught Altman less than nothing about humanity. The promise of automation was to do the mundane so human creativity can flourish. Instead, human creativity is demeaned as mundane so Big Tech's machines can flourish.

Let that be their world. It need not be ours.

6

—

Mindless Robots Can't Be Heroes

ALGORITHMS DON'T INSPIRE

We all know the story of Rosa Parks: How she was forced to sit in the "colored section," which occupied the middle rows of a segregated bus, because the front rows were reserved exclusively for Whites. How she couldn't even comfortably rest in that seat at the end of her workday, because the rights of Black passengers to occupy any seat on the bus were contingent on whether a White person, even one boarding later, was able to find a seat too. And how she changed the course of history when she refused the driver's instruction to vacate her seat and move to the back of the bus.

Rosa Parks became a hero by refusing to comply with a racist law. She questioned the moral sense of the police officers who showed up, asking, "Why do you push us around?" The police didn't demonstrate much moral depth when one answered, "I don't know, but the law is the law and you're under arrest."[1] A massive boycott of public buses by Black folks in Montgomery, Alabama, followed this incident, which ultimately led the US Supreme Court to finally outlaw racial segregation on transport.

What can we learn from her story? If one is an algorithmic supremacist, not much. Parks's behavior is seen as predetermined by a combination of biological and environmental factors. Just as the cop was mindlessly following a script that he didn't understand, Parks too was playing out the inevitable outcome of her internal neural processes, her unique upbringing, and the stressors of discrimination in that moment. But for artful thinkers, there's a lot to be inspired by. We can learn from her explanations, as she explicitly declared her defiance to be an intentional act: "I was not tired physically, or no more tired than I usually was at the end of a working day. I was not old, although some people have an image of me as being old then. I was 42. No, the only tired I was, was tired of giving in."[2]

To determinists, Parks may have thought she was acting freely, but all biological beings are automatons, lacking agency and simply going through the preprogrammed motions of life. But to us, the artful who see something special in the human experience, Rosa Parks made a choice, of her own volition, and her explanation moved people to see the justness in her cause. In fact, before her action, two other Black women had previously been arrested for similar resistance to racist norms on buses in Montgomery. But community leaders and civil rights advocates did not think these women would likely inspire others. There was something uniquely compelling in the story of Parks that made a diverse swath of humanity view her as heroic.[3]

Algorithms don't inspire. Brave choices, and the idiosyncratic way we defend those choices and widen sympathy, is are how we bring about social change. Algorithmic supremacists sell us the manipulative idea that the human brain, despite its literally mind-boggling complexity, should be viewed as a rather simple computational machine. But, by tapping into artful intelligence, we create an alternate reality. To stay valuable, relevant, productive, and free in the coming AI-dominated world, thinking with **BEAM** is what will ground us. But rethinking the process of cognition is not enough. We are social, embodied beings in need of meaning, connection, and wonder. Defending our **VICE**—volition, intent, choice, and explanations—is how we will not only get through the day but also create, innovate, and inspire the people around us.

As with our exposition of **BEAM**, choosing the acronym **VICE** is intentional and seeks to play with the term's more common usage. If we are entering a period where adopting algorithmic supremacy is widely perceived as virtuous, then those who oppose the paradigm will need to embrace some counter-traits. What have come to be viewed as vices in the minds of AI enablers must be defended by the artful. There should be no shame in loudly embracing the historically loaded language of vice in our mantras.

Let's break down the core components of defending your **VICE**. Big Tech's elite believe that the human sense of free will is a delusion. Their faith in determinism is a core principle informing how they choose to live and why they are so keen to be freed from biology. We build lives defending space for **volition**, acting as if we are free to choose, and believing that the future remains ours to shape. Keeping our lives artful demands effortful mindfulness to ensure that what we do, what we hope to achieve from **BEAM** thinking, always comes with a willful **intent**.

As human engineers do their best to force-feed us their data, limit our choices, and manipulate our rationality by muting embodied sensations, we can trust our bodily instincts, be irrational/illogical/emotional, and stay optimistic, because despite the horror stories, humans are still coming out on top. We are battling the misinformation, shining a light on biased algorithms, discovering the misrepresentation of autonomous vehicles, giving voice to the workers behind seemingly mechanized products, naming herding efforts, and pointing out deceptive designs—all the while still using technology that supports our efforts to think with **BEAM** and defend our **VICE**. Every one of these instances is a tale of a human outsmarting AI.

How to Be Free

A frightening number of engineers and AI thought leaders believe that our perception of **volition** is an illusion. With this certainty, they build tools designed to surveil, limit choice, and control. I imagine they justify the exploitation with the rationale that it is less problematic for a new puppeteer to take over the strings since the puppet has already been tied to the control. But it's interesting that those who see themselves as victims of their biology choose to then victimize their biologically based peers further by subjecting them to a new layer of engineering. Artful intelligence tells you to be free. But how do we know when we are free? What should we be mindful of?

Philosopher Derk Pereboom draws attention to four possible types of free will.[4] First, ask yourself if, within the situation you face, you have the power to create alternative possibilities. Are you free, in that moment, to take a specific course of action? Are you equally free to choose to do nothing? Part of artful living is that when we act, or when we refrain from acting, it is a willful consequence of our independent and unique **BEAM** processes.

When Dr. Block decided to get up on a table and dance in religious ecstasy, it was because he elected to go with the bodily sensations he felt in that moment, decided that the table was an appropriate resource in his current physical environment to express this ecstasy, and knew that what he was doing was itself a way of thinking about life. He could have refrained from dancing on the table, instead maybe clapping and singing while seated.

Dr. Block had awareness in the moments prior to taking to the table that there were alternative possibilities, different courses of action available to him. His near future could have unfolded otherwise. But he created a path that his **BEAM** sensed would bring him joy and meaning. In the aftermath, he laughingly defended this **VICE** to me and the other uptight onlookers who had chosen some of those alternate paths for ourselves. We too had access to free will and were using it to shape what was, admittedly, a less joyous future. When you dance, can you laugh at yourself? When you choose not to dance, are you sure there was a choice? Or was the weight of your heart too heavy to be free? These are questions artful thinking demands you wrestle with.

Next, are you free to act without those actions being predetermined by causes beyond your control? Dr. Block never used the language of "the spirit compelled me!" in defending his **VICE**. Nor, despite the consistent presence of vodka at these Hasidic gatherings, did he ever describe the table dancing as the inevitable consequence of his drinking. Which is not to say that there aren't moments when we may be compelled to act by causes out of our control, be they physical or metaphysical spirits. The human engineers of algorithmic supremacy are working hard to ensure that more of us find ourselves in increasingly frequent situations where we have been externally compelled to act. But those decisions are not a consequence of thinking with **BEAM** and not the type of actions we would artfully defend. There is no cognitive error in recognizing that you are being manipulated by the algorithm on your phone. The error occurs in staying willfully blind to the fact that you were never truly free to choose.

Third, are you free to act in ways that are responsive to reason? For the artful, "rationally" is a loaded term that has been co-opted by algorithmic supremacists. Transhumanists, rationalists, and longtermists aspire to program humanity in ways that compel excessive amounts of a certain type of rationality. What is more important is that what we do with our free will is guided by some sort of **BEAM**-inspired reason, like a bodily sense, an environmental inspiration, or a principled goal. The justifications we offer need to be our own, in that we need to have come to them ourselves. If you only parrot talking points, if a stranger can predict your stance on every issue based on your perceived identity, then you aren't truly free. To be a thinker who is meaningfully free is to regularly change your mind.

Finally, are you free to control yourself? Artful living involves defending morally motivated actions. If what we do can't be defended, if we don't have persuasive words of justification, then we hold ourselves accountable for the consequences. As discussed in the opening to this chapter, Rosa Parks managed to change US history by taking action that increasing numbers of people viewed as moral. She took care to express that it was a freely exercised moral move, not born of fatigue or some other physical inability to vacate the seat. She owned her decision and all the consequences. The arresting officer also saw it as a move rooted in free will but was uninterested in hearing the moral content; to him, all that mattered was that the law was on the other side of this issue.

Rationalists like neuroscientist Sam Harris make the peculiar argument that criminals do not have the free will to control themselves. Yet nonetheless, Harris wants them to be held morally accountable for their unfree acts. To him, criminals are always victims of their biological or environmental circumstances, but punishment is nevertheless appropriate because the desired effect is an increase in "the well-being of all concerned."[5] Sharing a true story about a pair of repugnant murderers and rapists, Harris confesses that "as sickening as I find their behavior, I have to admit that if I were to trade places with one of these men, atom for atom, I would *be* him: There is no extra part of me that could decide to see the world differently or to resist the impulse to victimize other people."[6] Society still needs to be protected from people programmed to act that way, and they need to be placed in some sort of institution for their own well-being since they are incapable of not causing harm. To me, though, this hyper-rational type of reasoning doesn't create space for meaningful morality. What Harris offers is a practical solution to a social problem, an algorithm for order, not the basis of a moral system.

To be clear, while I advocate for believing in free will, I am not proposing that if we answer "yes!" to these four questions we can be as sure of our freedom as we are of the ability to take a deep breath. The call of artful thinking is to *defend* your **VICE**, which is to say, make the choice to believe you can willfully make choices. Algorithmic supremacists want to build a new social order rooted in certainty. We, in turn, are here to challenge their claims to certainty, ever mindful of the enduring mysteries that penetrate every segment of our lived experience. Be mindful enough to ask yourself why you are shying away from the dance floor, why you refuse to change your mind, why you can't put down your phone and be your own biggest cheerleader of your efforts to change—even when there isn't the certainty that change will happen.

Philosopher Mark Balaguer describes exercising free will as engaging in a moment of conscious decision-making.[7] But what are we accessing when immersed in this fleeting state of reflective consciousness? What resources are our body and/or mind tapping into? A spiritual person might answer that we are utilizing metaphysical resources, connecting to our soul or higher self. But to a rationalist, the only explanation is that these moments, or more correctly, all moments, are wholly physical. In the materialistic, scientific view of humans, we don't access, deliberately or unconsciously, a nonphysical reality. Every seemingly conscious decision is the result of a physical process, specifically a neural event.

What triggers these neurons? Factors beyond our control—our genetic makeup; chemicals, like adrenaline, dopamine, or serotonin, surging through our body at that moment; our childhood experiences that shape our impulses, etc. To them, Rosa Parks had no choice in that moment but to refuse to move to the back of the bus. The trauma of her childhood, the effects of racism, the elements of her personality that so endeared her to people, made it so she had access to only one response in that moment, despite her claims that the resistance was intentional.

Except, this wasn't the first time she was told to move to the back of the bus. It wasn't the first time she experienced racism. It wasn't the first time she thought of the possibility of resistance. What evidence do rationalists have to suggest that there was something absolutely unique about the racist encounter on December 1, 1955, that triggered a preprogrammed response that wasn't triggered during seemingly similar encounters on other dates? In that moment, Parks faced a quandary she had faced before. And the nature of that quandary is the best evidence for the possibility of free will.

Balaguer calls it a "torn decision," making a conscious choice when you have multiple options that appear to be more or less equal in their attractiveness. In that scenario, one would feel completely unsure of what to do. Nonetheless, if it is our intention to proceed with life, a decision needs to be made while in this state of feeling "torn." Balaguer's "torn" decision-making has strong echoes to philosopher Jacques Derrida's teachings on undecidability, an approach to thinking that has had a massive influence on my research in ethics.[8]

Derrida believes we make ethical decisions when we feel an urgent call to get something done. The voice calling out to us may be external, a person in need whose cry stirs something inside us. Or the voice may originate from within, born of a massive sense of responsibility. Either way, there needs to be a real-life situation, something happening in the physical world, that we can affect with our actions, like Parks encountered on the bus. Morality never begins in the realm of contemplation, when we're comfortably schmoozing. The responsibility for reflection only arises on account of the pressing need to get something done. Think about that: armchair moralizing is not artful. Until we think with action, until we do something real, we remain locked in our heads, away from the moral realm.

Derrida explains that in these situations, whether one calls them being "torn" or "undecidable," the main challenge is that "often, particularly when it comes to ethical decisions, there will be two competing choices that seem to have an equal amount of truth value, and the moral actor will not have access to any resources that will aid him or her in objectively justifying one decision over another."[9] Rosa Parks must have been torn, must have faced undecidability, with every racist encounter.

On the one hand, there would be the voice inside her demanding justice, the part of her that was tired of being victimized by racists. On the other hand, she knew there would be a heavy price to pay for resisting. She knew she would be arrested, would lose her job, and would receive death threats, and indeed, all these painful consequences did materialize. No one would have faulted her for moving to the back of the bus that day, as she had in the past. There were good reasons to resist. There were equally good reasons to submit. Nothing in that moment provided indisputable solid evidence for making one choice over the other, which made that moment an opportunity to exercise free will. And that's why she is celebrated.

For better or worse, our freedom has limits; we are not constantly making torn decisions. When we turn off our brains to watch a mindless sitcom, or are otherwise on autopilot, we're not facing torn decisions. Maybe most of the day we don't exercise free will. And maybe it's only in those rare moments of being torn that we break out of our algorithmic programming, snap out of our stupor, and act artfully, even if just for a beat or two, to make a conscious choice of what to do next.

But you know what? That's enough...for me, it's enough to know that in those precious seconds when we consciously make torn decisions, we are truly free. The actions Rosa Parks took on that fateful day were not predetermined. Those were choices, made in those moments, that allowed her to taste freedom and be motivated to fight for greater freedoms. There is simply no algorithm for experiences like that.

Balaguer adds that to the scientists and engineers who still want to argue against the possibility of free will ever occurring in anyone's lived experience, we can respond that maybe our torn decisions are *sometimes* influenced by unconscious factors that are out of our control. But there is simply no evidence to support the assertion that our torn decisions are *always* caused by unconscious factors. We don't have any solid, indisputable, rational reason to back away from defending the existence of free will.

As the algorithmic supremacists commit to certainty, choose to live in the moments that undermine certainty. When the algorithm tells you to deny care, deny hope, deny giving a person who stands before you with an open heart a chance to be heard...walk away from your screen. Think about the implications. Turn the decision into a dilemma. Move from a state of certainty to a state of being torn. The artful always try to maximize opportunities for making torn decisions, so use your tools to enhance, not minimize, questioning and uncertainty.

If you want a practical way to find more freedom, upgrade your attentional flexibility—that's the ability to mindfully shift your attention between different objects or levels of focus.[10] What makes you stuck? It's seeing the same things wherever you look, realizing that your attention is always captured by the same few variables, every day, as you undertake the same repetitive activities at work and at home. It's not that routines are problematic. Routines are good. It's operating in habit mode that prevents us from exercising free will.

You can have a routine, like eating all your meals at certain times or going for an evening walk. But within the routine, try to shift the focus of your attention away from the habitual. Peer ahead, not because there's a grand new truth on the horizon, but because there's something up ahead that is equally compelling to what you are used to focusing on. That may be enough to make you torn. Turn your gaze not because you are some combination of jittery, distractible, or unable to focus but just because it might make you more free.

With the Best of Intentions

The Buddhist monk and writer Thich Nhat Hanh taught "good intentions are not enough; you have to be artful... Mindful living is an art, and each of us has to train to be an artist."[11] Building a human-centric future requires **intent** in support of meaningful presence. Intentions are motivators, not just in initiating action but also in sustaining the action until we have completed our intended task.[12] Elisabeth Pacherie identifies three different levels of intentions.[13] Each type of intention serves a different purpose, and we can hold multiple intentions at once, especially as we are transitioning between the different levels.

Let's reframe our soup-making example as a story of intention. I shared earlier that my grandmother would alternate between *kneidlach* (matzo balls), *lokshen* (noodles), and *kreplach* (dumplings), depending on the Jewish holiday season. So, for example, as the winter thaw would begin and family discussions turned to who would host the *Pesach seder* (Passover feast), my grandmother would start thinking about the vast quantities of egg noodles she would be forming from scratch, by hand, to feed her guests. This category of intent involves high-level planning for the future. Scientists call it "distal intent," which acts as a coordinator and prompter of practical reasoning about means and plans that we are not going to take for a while but need to start thinking about well in advance.

Now consider our Redditor. He was deeply immersed in his online world but knew that his body was calling out to him and he would have to feed himself at some point in the day. So, he asked a bot to help with his planning. This type of intent is still high-level but is present-directed or "proximal." These intentions are responsible for conscious guidance and monitoring to ensure that imagined actions unfold appropriately in present time. For my grandmother, "proximal" intentions might have been activated the night before the Passover meal, when she decided the time was right to get to work on the noodles and start prepping for the meal. For her, this mindset would bridge the gap between distal intentions and the actual performance of the action.

The final type of intentions are low-level "motor intentions," responsible for unconscious guidance and monitoring. They are formed at the time of the action and direct the details, like the intricate variety of muscle movements required to get my grandmother's hands to crack the eggs, stir the mixture, pour it into a pan, flip it out, roll it up, and cut the individual noodles. It is the intent that controls the physical processes we're not fully aware of.

Imagine that as she's making her *lokshen* the phone rings. Her motor intentions are jolted away from actualizing the proximal intentions as she answers the phone. It's a relative who will be coming to the meal tomorrow and shares that they can't wait to indulge in my grandmother's *kneidlach* (matzo balls). Suddenly, her distal intentions are misaligned with her proximal intent. Her best-laid plans cannot be implemented without drawing the wrath of a relation. Her proximal intentions are altered, and her motor intentions send out new instructions to the muscles in her fingers as she takes out the matzo meal and quickly switches course.

Philosopher Michael Bratman describes intentions as "terminators of practical reasoning."[14] In other words, when an intention kicks in, it becomes a shutoff valve to the deliberative functions of our cognitive process. Once my grandmother made the willful choice to declare internally the intent of making *lokshen* with her chicken soup, unless important new information was acquired between that moment and when she started the cooking process, the intention resists reconsideration. But when the phone rang, everything changed, as her intentions now served to coordinate her actions with the actions of others. In this way, defending **VICE** can become a communal activity. Algorithmic engineers want to remove intentionality from the equation to isolate us and subject us to tuning. But mindfully claiming our intentions gives us a path to connection, weakening the power of hostile actors seeking to manipulate us.

We may think that changing our minds is a simple thing, but it's not. There are numerous cognitive processes in play. Before we act, our intentions (1) terminate further internal deliberations about what ends we should be seeking, (2) start a process of practical reasoning about the means we will need to now undertake to achieve those ends, and (3) start coordinating a plan to action. Our intentions then (4) initiate action and (5) sustain the action, all while (6) guiding and (7) monitoring both conscious and unconscious supporting behaviors that will lead us to eventually completing the task.

Changing our minds is a significant interruption, requiring a new process. We just said that true freedom is in the ability to change our minds but look at the myriad of internal processes working against us. Human engineers know this process well and try to use it to control us. We need to understand our intentionality so we can build defenses against their intrusions.

We are most alive, most human, when constructing new realities by willfully doing things in the world, sometimes because we changed our minds, other times because we shifted our intentions, and occasionally from finally making a move after being torn. Algorithmic supremacists undermine the terminating function of our intentions by handing us devices that give them constant access to our attention, continuously feeding us prompts we may mistake as relevant information. The IoT is at its most devious when gathering ingredients in our house can initiate a system of prompting like "Are you sure you want to make a soup? Here's an ad for Uber Eats...the soup will arrive much faster than you can make it." Holding on to our intentions is no easy task in the age of AI.

Alva Noë takes a big-picture look at intentionality, using the phenomenon of reading[15] as an example of the magic behind artful intentions. As you read this book, ask yourself, where are you directing your intentions right now? It's not at the individual letters, or even the particular words, that are presented to you on the page. In fact, if in reading this or the previous line you now choose to willfully focus your attention on the letters, it's p r o b a b l y going to disrupt your ability to keep reading.

Noë recognizes that we are landing on something of a paradox here. You can't read this paragraph if you don't see the individual letters. But if your intention shifts to really seeing the letters, you won't be able to read the paragraph. Rabbi Tzvi Freeman beautifully states, "a paradox is not a blunder of logic. It is a discovery of wonder… What appears to us as an irresolvable conflict of two aspects of reality forces us to see a higher reality."[16] Add that perspective to your cognitive toolbox. Paradox breaks the algorithm but fires us up. If you want to stay productive in the coming era, play with paradox, let it turn on your sense of wonder, and expand the realm of possibility open to you.

It may be the case that the soulless algorithmic supremacist chooses to, I would say, willfully blind themselves to the awe-inspiring wonder of human intentionality. AI built on the neural net architecture "writes" by "looking" at the individual letters and running probability assessments to try to predict what letter should come next. It "paints" by looking at an individual pixel and running a similar algorithm. That's not what we do when writing or painting.

What makes reading possible is the possession of a very complicated set of skills. Reading only takes place against the background of our culture, our expectations, our history, and a myriad of factors that support the possibility of intentionality on both a conscious and unconscious level. Noë describes perceptual consciousness as "a species of intentional directedness" for which "one needs skills of access." Our minds, which include both brain and body, do not passively receive information. We are actively directing our intentions toward a target that we are curious about. It's **BEAM** thinking—using body and environment, doing things to make sense of what we are perceiving. And, not only do we spend a lifetime developing these skills, but we also reap the evolutionary benefit of our ancestors' efforts as well.

To engage our intentions is to establish a relationship with something. "Perceiving is exploring the world. It is a temporally extended activity. What we call seeing the apple is just an episode of exploration… We enact the perceptual world by skillful exploration." I keep coming back to the word "magic" in describing the revelations of natural intelligence because I think each new level of depth we unfurl can only be described as such.

Intentions are wondrous, relationship-building displays of skill. They need to be a passionately defended **VICE** against the negating efforts of algorithmic supremacists. Recall our discussion of some original biblical God tech in Chapter Three. The builders of Babel were not in awe of the magic in their intentions. Despite the fact it was so soon after the devastating Flood, they nonetheless had developed the ability to willfully connect with every other living human being. That is magic. They were able to map out and execute a plan for constructing an unprecedented technological advancement—magic. They were able to sustain those intentions so far as to require a divine intervention to put a stop to the project. If needing a god to counter your powers isn't magic, I don't know what is.

There was so much skilled human power on display in the intentions of the Babel tower builders. But what undid the magic was the decision to direct their intents toward antihuman ends. In contrast, Eve, one of the first human beings, lacking the benefit of history, used her intentions to test the intentions of God. To use the language of today's algorithmic supremacists, Eve did not engage with the God-like tech to see if it was in *alignment* with her intentions. She engaged in the tech with *ethical* intentions, wanting to break the rules to build a more compassionate, human-centric future. She defended her **VICE** and changed the course of the world for the better.

Choose More

When we live as if we are free, mindful of the power that emerges from willfully directing our intentions, then world-altering meaning is imbued in every **choice** we make. Algorithmic supremacists see the constraints of our DNA programming as absolute, such that the process behind our choices is not materially different from the computational choices of an AI. We grant that there are boundaries to our abilities, but there's still enough room within those boundaries for a very meaningful life. Frankly, if the determinists are right, then life is not all that interesting. And artful thinking is designed to keep life interesting.

One of the most empowering recent findings in this regard for the Team Human camp is that when faced with choosing between an option that leads to a set action (the end of the road) or an option that leads to a subsequent choice, humans tend to prefer the option providing further choice.[17] In **BEAM** decision-making, this research implies that we should always be choosing paths that offer more future choices. The most defensible **VICE** is the one that keeps our options open.

As explored way back in Chapter One, the preference of algorithmic choice is to reach the end of the algorithm as quickly and efficiently as possible. The algorithmic mindset wants the fewest steps possible, and the algorithm has served its purpose when the opportunity to choose has come to an end. We, in contrast, are wired to look for the chance to keep choosing. Making choices is life affirming. We don't want it to end. Living by the principle of "always more choice" is the best strategic response to the herding efforts of surveillance capitalists. To constantly be on the lookout for paths leading to more choice may be the best way to prevent being herded and controlled by algorithmic tools. Choose your model of the world, your choice of language, your drive to keep choosing the paths that lead to more choice so that you map out a full life.

Nobody Says It Better

Finally, put the highest value in the **explanations** offered by fellow artful humans who justify their choices. Prize heartfelt surprise over the predictable computational output of algorithms. Mike Mignola, the creator of the comic book character Hellboy, explained to me once that his drawing choices are made in a fully mindful state that lots of other artists may be technically better at drawing than he is. But he would look at their work and have no sense of who they are. He saw limited artful intelligence in these drawings. While any number of people can draw Hell, with Mignola's Hell we are looking at the inside of his head—it's uniquely him. As he said: "It's not better than anyone else, but no one could put those things together the way I put them together."[18]

Artful intelligence builds on the uniquely human skill for explaining our processes so that we, and others, can fully understand them, replicate them, and improve on them. An algorithm can't explain what it "learned," how it is "learning," or even why it came to the decision it did. Even the programmers of a system have no clue what their AI is learning or why. The holy grail of future AI research is in the demand for explainability, the capacity to express why an AI system reached a particular decision, recommendation, or prediction. It's a chalice we can drink from now.

Angus Fletcher, a professor of story science (a job title I'm very jealous of), believes that our powers to explain are tied to "storythinking," which is an ability for "thinking in actions,"[19] like trying to figure out why something happened, guessing what will happen next, or imagining how things could have happened differently. Information is easier to encode when wrapped in context, and stories are optimal vehicles for providing context, particularly when they are personal.[20]

An ability to storythink will always give us the edge over AI. Fletcher asserts that "our personal physical, emotional, and intellectual growth is accelerated by empowering the storythinking of the people around us." Parks's actions were important precisely because they triggered a process of transformation that included stimulating the storythinking of so many people in Alabama who then took their own actions.

Noë muses that the power of our explanations comes from the fact that we are something like pieces of art, in that our true nature can never be known but instead "unfolds in the activity of working to know and see."[21] We try to understand ourselves, much like we try to understand a piece of art, when we attempt to explain what we see, feel, or need.

"There are no tests we can run, no brain scans we can study, to get to the bottom of ourselves... We are works in progress."[22]

Often, we ourselves don't fully understand the motivations behind our choices until we are forced to talk out and explain ourselves to others. Determinists aren't wrong in noting that some of our choices are made without will or deliberation. But the artful are committed to the search for meaning and strive to understand our choices when possible. Feeling the need to explain ourselves is a great tool to help us more fully understand when we are, or are not, fully in control.

Meaning, like consciousness, is a concept that scientific rationalists struggle with. Algorithmic supremacists love to work with information because it can be explored through a mathematical framework. As we noted at the beginning of the book, despite Leibniz's best effort, meaning can't be quantified in a mathematical equation. Meaning is "inherently historical, relational, contextual, qualitative, even subjective... That makes the concept of meaning at best awkward and at worst fatally suspect from a scientific perspective. Yet none of that implies it isn't real."[23] Artificial intelligence is designed to compute information. Artful intelligence enhances our abilities to discover and create meaning, to find meaning in what we and other people do.

And just as we ended our development of thinking with **BEAM** by citing a scientific breakthrough, we will close our discussion of **VICE** with an equally paradigm-shattering scientific discovery. Researchers announced in July 2023 that our brain waves synchronize when we interact with other people.[24] Let the full impact of that finding rock your world. We have been arguing that whether validated by scientific research or not, hopeful individuals should choose the artful paradigm of defending free will, intentionality, and the surprise of heterogeneous explanations to redirect society in a more human-centric direction. Is there a wilder explanation for social connection than the syncing of brain waves?

The research comes from a relatively new stream of neuroscience that looks at collective effects. It seems that when people share an experience, there is strong evidence of their brain waves synchronizing. These scientists report:

> *Neurons in corresponding locations of the different brains fire at the same time, creating matching patterns, like dancers moving together. Auditory and visual areas respond to shape, sound, and movement in similar ways, whereas higher-order brain areas seem to behave similarly during more challenging tasks such as making meaning out of something seen or heard. The experience of "being on the same wavelength" as another person is real, and it is visible in the activity of the brain.*[25]

To anyone living artfully, this should come as no surprise at all. Wilco's Nels Cline explained to me that the magic of playing music in the setting of live performance is the opportunity to create not only with his fellow performers but together with the audience as well. He observes that something physiological must be happening when we all gather in the space of a concert hall or theater. Nels sees that in those moments all who have gathered are sharing something…it's not just escapism, or a rally; it's something more. Nels described it to me as an experience leading to the feeling that we are all coming out transformed at the end of the night, feeling cleansed.

Collective neuroscience now backs up what Nels claims (not that it needed scientific support to be a compelling explanation). Neural waves of people attending a concert matched those of the performers, with the greater the synchrony in brain waves, the higher the expressed enjoyment by all who were there. When students are engaged, their brain waves align with those of the teacher. Couples and close friends show higher degrees of brain synchrony than other acquaintances. This explains why we don't always "click" with certain folks.

When I chatted with Nels about this book project, and how his words continued to resonate, he lamented the feeling that he's working against the cultural trends, seeing it as "troubling in so many areas as we drift into this seemingly inevitable frontier. For myself, I feel pretty much like an old theater person or traveling circus performer or 'hoofer,' dedicating myself rather resignedly yet steadfastly and consciously to the dying art of 'live,' of the immediate and often unscripted. But not in protest or opposition, really. It's just what I like and a lot of what I know. And I guess I want to represent this worldview without standing on a soapbox, but by example. Let's at least try to imagine the future being better than the many dystopian imaginings."

Indeed, there is cause for hope and optimism.

PART III

ARTFUL MODELS

7

Blue Skies Are Better than the Metaverse

WHAT DOES AN ARTIST KNOW?

I saw Jane's Addiction perform for, what I thought would be, the last time in the summer of 1991, when they brought an all-day, multi-band, multi-genre, farewell traveling festival called Lollapalooza to my hometown of Toronto. After nine hours of sets that in any other context would have been showstoppers on their own, Jane's took the stage.

The first sound heard rolling over the roar of the crowd was Eric Avery's distinctive bass riff, providing the melody that would carry the opening song, "Up the Beach," followed by a psychedelic wash of Dave Navarro's guitar joined by Stephen Perkins's intense tribal drumming. Singer Perry Farrell moved to the front of a performance space adorned with flowers, plants, statues, and mysterious icons, holding a bottle of wine, wearing a mischievous grin, vocalizing with the band, and sharing only two words: "I'm home."

We ended the last chapter talking about the neuroscience of collective effects. Every person gathered on that August day was experiencing synced brain waves, with neurons firing at the same time, imprinting the specialness of the moment in both our individual and collective memories. That show had what Abraham Joshua Heschel calls a "soul-shattering"[1] effect on me—an encounter that my intellect could simply not fully process. Mindful of what being in that artful space did to my brain, I was left wanting more. So, when Perry and Stephen announced a new project, Porno for Pyros, a year later, I made my way to LA to again be in the same room as these uniquely creative people and the artful folks who gather with them.

Perry is the master of throwing a party, famous for finding novel ways to create physical environments that fully engage all the embodied senses. His planning focuses on the needs of the people coming out to the space to be with him, very much reminiscent of how Bill Graham used to promote the Grateful Dead, offering free fruit, water, and other comforts, and in opposition to how corporate monolith Live Nation presents its shows. For decades, I found soulful inspiration in Perry's music and art, and the unique cultural gatherings he hosts. So, in 2023, when I got word that PfP would be reuniting to play a special event with the original lineup for the first time in twenty-six years, a much older me did not hesitate to make the pilgrimage to LA once more.

The venue Perry and his wife, Etty, chose to host this event as part of their *Heaven after Dark* party series, the Belasco Theatre, is gorgeous. We wandered through themed rooms upstairs and down, indoors, and even on a terrace outside. There were dancers, stilt walkers, fire breathers, and costumed beautiful people roaming the venue, giving out flowers, leis, smiles, and energy to assure that the gathered brain waves were in sync before a note of music had even been played.

When PfP finally took the stage, the crowd was well primed. The MC for the night, equipped with angel wings, spun around making lewd references as she let us know the moment we were waiting decades for had come. Stephen, bassist Martyn LeNoble, and guitarist Pete DiStefano took the stage first, beaming and pointing at each other, allowing each one to have a few breaths taking in the adulation. Etty and another dancer joined at side stage. Perry hopped on last, dressed simply with a white tank top over a striped one, and they launched into the band's eponymous track, with every band member joyfully mouthing the lyrics.

I cannot overstate the role Perry Farrell has played in shaping my artful intelligence. It feels right to start this section of conversations with someone who identifies as an artist in the classic sense. Perry is now in his sixties and no less fearless than he was in his youth. I'm aware that artful preferences are idiosyncratic, and our unique lived experiences lead us to appreciate different types of sights and sounds. So, if it happens that you don't know who Perry is because your sonic curiosities led you elsewhere, please refrain from Googling or Binging him until after reading this chapter. I want you to meet a creative visionary who may not be well captured by a Wikipedia entry or a random YouTube video. The Perry Farrell we will be hearing from was generous enough to show many sides of himself in these frank and open exchanges: musical innovator, of course, but also entrepreneur, teacher, idealist, husband, Jew, elder, spiritual advocate, and more. Get to know him through the intentions, choices, and explanations he's offering here.

Perry entered the room with a huge warm smile, looking slender but strong, making deliberate eye contact, and simply beaming positive energy. "Hello, my friend," he offered in a comforting voice. "Nice to see you again." I consciously let his sincere greeting sink in, which is not an instinctual response for me. It's part of the work I need to do to become more artful. If volition and choice are the defining characteristics of who we are, then these little moments of social exchange take on an outsized importance. They need to be felt with the full body, not barely heard and then quickly dismissed and forgotten.

Algorithmic supremacists want a society conditioned for the rote greetings of automated tools, like the synthetic reception provided by a QR code on the table at a restaurant or bar, where your first contact used to be a host/hostess. Or in the way social interactions are minimized in retail environments as cashiers and greeters are replaced with screens and self-checkout booths, with humans retained only as security to monitor shoppers at a distance. When the social aspect of public spaces is removed, it becomes easier to forget what it once felt like to authentically connect with a person, even for the briefest of moments, before moving on to the business at hand.

We are being primed for an economic future with limited human-to-human interaction. Start mindfully appreciating real moments of human connection when they occur and take the time to recognize them as such. Knowing how brain waves sync in social situations should incentivize us to spend a minute allowing the syncing to take place, feeling it, knowing it, appreciating it. These are the simplest of exercises for adjusting our thinking back to a more human direction amid the algorithmic onslaught.

In my conversation with him, Perry intuitively landed on one of this book's central messages. The ideological battle of algorithmic supremacy versus artful intelligence is the fight between those wedded to determinism and those celebrating free will and choice. Perry observed, "The discussion that we're having today is really on the role of freedom of choice, and how to respond to the people that believe we don't have freedom of choice. They're looking at life rationally or scientifically… which is fine." Perry continues, "The world was built on choice. Some people, in the circumstances of their daily lives, might feel like they don't have the power of choice. But I'm here to tell you that we all have freedom of choice. If one were to think only with reason, one might conclude that we don't have choice. But being that we are souls, clothed within bodies, here on earth, we are given the freedom to choose between good and evil. And through those choices comes love and awe."

To choose the path of choice is not to be under the delusion that all options are always available. In much of our daily lives, the opportunity to be fully free is limited. As we saw in our discussion of being torn, sometimes we act on impulse or bias or biological need, without exercising free will or conscious intention. But the prescriptive move is holding tight to the knowledge that the freedom to choose does emerge on occasion. And when it does, make the choice that is most true to you.

Nowhere are the world-building implications of freedom of choice more apparent than when we are faced with moral/spiritual choices. We construct new realities when choosing to redefine meaning for ourselves, or welcome new social connections, or even by opening up to the feeling of wonder. This ability, to build new worlds simply by choosing more openness, is the quintessential human experience. It engages the head, the heart, and the soul.

I was lucky to catch Perry during a time in his life that was filled with reflection, renewal, and rebuilding. He had made the choice to heal frayed relationships, intentionally choosing to try to shake off the jadedness and fatigue that can come with being a "been there, done that" creative icon. After decades of relational instability, at the time of our meeting Perry was back in the studio recording with Eric Avery, Jane's Addiction's original bassist, who had returned to the fold following years of absence. They were revisiting the creativity that started in their youth and was put on hold through middle age. Human creativity, feeling, and art, can change and grow over time. And these changes don't come from simply gathering more data or advancing further along a predetermined algorithmic path. Something else happens, and I wanted to hear Perry talk directly about the embodied feelings he was aware of at this point in the narrative of his artful life.

Art without Algorithms

Perry responded with humbleness and sincerity, "It is just so amazing that I got a chance to write this final chapter. I didn't get wiped off the earth.

I have a chance to complete it." As he spoke, his face shifted from the captivating smile that often animates his public visage to a more strained expression. I could tell that he was allowing himself to be vulnerable, that he wanted this truth to come out. "But I am concentrating so hard on finding the right words, the right poetry, the right delivery, the right sound... You know, it's song and psalm, right?"

Perry looked at me with intensity, explaining why the creative process this time around was different. "It's got to be just so. I don't want to say I'm struggling... But yeah, I am struggling...I'm struggling to find that balance. But that balance must be struck, because if I can pull this off, it's going to be fantastic. I want folks to go 'I'm with you, man!' That's what the struggle is," he said, punctuating the expression with a fist pump.

This emotional struggle exemplifies what it looks like when an artist is at work. It is the antithesis of the process being normalized by AI's algorithmic tools. AI art comes from prompts and probabilities, spewing out a variation of something already created. Perry works to find the balance between what he wants to create as an artist who has the burden of a cultural legacy, who has albums under his belt that literally inspired a generation of musicians, but who knows that his past work had the effect it did precisely because it presented something novel but unifying, speaking to the moment while invoking change.

Good art empowers choice because it touches (or shatters, as Heschel said) the soul. It's an experience that, while decidedly human, reaches us on a higher level than rational argumentation. For those uncomfortable with soul-speak (although, according to a 2023 survey, 83 percent of adults in the United States do embrace the language of having a soul[2]), we can reframe it in the vernacular of **BEAM/VICE**. Art is created by embodied beings as a method of transcending the limitations of their bodies, with the intention of impacting the sensations of other embodied folks. It is thinking by doing, in particular environments, that transcends the creative medium being employed (be it music, words, images, or movements) to inspire other actions, by other people, and reach into other environments. Art is intentional, a choice to communicate and explain an inner truth for which words or tools of rational persuasion prove absolutely insufficient.

It is why, as writer Laura Pitcher notes, AI art is "so cringe."[3] Lolita Cros, an art curator interviewed by Pitcher, allows for the possibility that the AI medium will eventually become accepted as a common tool for legitimate artists. The current soulless state of AI art is the fault of the people behind the algorithm. Cros explains, "If algorithms can be assembled by a thoughtful human with good ideas, then I can't wait to see the AI-generated work they'll come up with. The problem is when the humans creating those algorithms are uninteresting and uninterested in art." So long as algorithmic supremacists are boosting the tools, there is no hope. And that may be the case for a while, as all current AI tools are trained on stolen creative work. But future models will be trained differently. Perhaps then artful folks will be more interested in using the tools for environmental off-loading.

For now, self-respecting artists like Perry have little interest in AI. Perry tells me that the biggest shortcoming from his early experimentation with chatbots is the lack of soul behind the output. What Perry found in the "speech" of AI was the expressions of a programmer, dressing in extra layers of digital clothing, to hide and distort the human presence behind the algorithm. There's no beauty without a flash of soul. Perry advised, "If you want to write a song about the spirits on Earth, you need to have that experience."

Etty jumped into the conversation, sharing her take on bodily wisdom and artful expressions from the perspective of a dancer: "As you see, I talk a lot with my hands. I act out my words, and sometimes I would be talking with my friends, who are dancers as well, and I would say: 'Can't I just dance this to you? Because you're not understanding what I'm saying.'" This is a beautiful example of the **BEAM/VICE** interface. Sometimes the best explanation or, at least, the explanation born of the most authentic intent can only come through bodily action. Etty continues, "Baryshnikov once said something very poignant: 'Dance for me a minute, and I'll tell you who you are.' I feel that. I pick up a persona from how someone moves."

As Etty spoke, Perry was staring lovingly at his wife, and started riffing on some Kabbalistic ideas. In the Tanya of Rabbi Schneur Zalman of Liadi, a text Perry studies, our conscious thoughts, words, and actions are described as garments for the soul. "We are brought down [to the material world of action and] we're clothed within this garment…and the gifts [of thoughts, words, and actions that] some of us get, like Etty, the expression is beautiful…rhythmic…artful, and balanced." These expressions, whether you describe them in "soul" or **BEAM** language, are bright flashes of communication that pierce through the veil when we use our body to reveal an inner truth. Our bodies constantly communicate information, even for those of us who are not dancers, or not mindful of it happening. Perry confesses, "When I'm speaking about something that is from the Porno era, I have different movements. It's body language. I'm not mindful of it at all. Going back to our subject of AI, here's the bottom line for me: It doesn't know how something feels…so you really can't tell me how to deal with something I'm going through if you can't feel."

If you want to think like a human, learn more about the quirks of your own body language. Ask the people who are around you most what they have noticed. Lean in to those moments when you are mindful of the limitations of your words and realize that you are relying on gesture or other physical expressions to move a conversation forward. Remember, gestures are a key processing tool for your own mind. They're not just a way to communicate a signal to someone else. They're part of how we humans think. If, like Etty experiences, there are moments when you feel you need to dance for someone to get a message across, do it! Holding back is compromising your cognitive processes.

I See You

AI art is significantly more than cringe-worthy. It is designed to discourage us from thinking in human ways, separating us from other humans in the creative process. There's no need to gesture, dance, or make eye contact with algorithmic systems. Eryk Salvaggio, an interdisciplinary design researcher, observes that "AI severs the community entirely. We talk to the AI, make art with the AI. AI is Web 3.0, a simulation of the web based on the ghost of interactions past. The tech industry seems likely to tighten the boundaries of these online systems so that we don't need anyone else to do the things we love. We will type for the machine that surveils us, share with simulated audiences."[4] The worst part of the initial AI rollout, considered a feature by the engineers of control but a bug for the human-centric, is that it is antithetical to creating new social worlds. We create with, for, and from the AI, alone and isolated.

We get to know ourselves, and our intentions, better when we are called on by others to explain to them what we are doing. We think better in shared spaces where our brain waves sync. I traveled great distances (more than once) not only to see Perry perform but also to be in the same room as the myriad of creative folks in his orbit. And Perry is always mindful of the role community plays in his creativity: "You know, with Jane's Addiction, and all the people around us, we were very lucky to have this art collective to make discoveries with. Our career, our line of work, is to make these beautiful gatherings for people to witness each other. We're basically building cultures. We all get together to listen to music that we like, and check out each other's style, and listen to each other's points of view. Being around others, expressing and exchanging love, that's the way to go. That is what I've made a career of."

To a certain extent, we have the power to choose our surroundings and who we vibe with, but for most of us these are arenas bearing heavy constraints. It's not easy to find new spaces or new people to be with. But one prescription that any of us can enact at any time, as mentioned in the last chapter, is to enhance our attentional flexibility. Stop focusing on the same details, the same annoyances, the same backgrounds. Tap into new intentions. Be surprised by the resources our environments have been offering all along but that we hadn't noticed because our attention was focused elsewhere. Once you embrace this practice, you'll be shocked at how much you have been missing, how repetitive your focus has been day in and day out.

"I cannot look at a computer or a cell phone for even as much as an hour without my eyes starting to hurt, and when my eyes start to hurt, my brain starts to hurt, and then my whole entire body starts to ache," Perry says. But he wants us to know that there is a fix: "Do you know how you can heal your eyes? Look up at the beautiful blue sky. Don't stare at the sun—look away from the sun—but look at an area in the sky that's blue." As he says this, he and Etty look out a window and take in a very bright California sky for a beat.

This is what it means to expand attentional flexibility. Make a more concerted effort to pay attention to, and be curious about, immediate surroundings. This tactic is especially important when you find yourself deeply consumed by a cognitive challenge, and your brain is telling you that looking at the physical space you are in would, in fact, be a "distraction." Resist that algorithm.

What happens after we let in the blue of the sky? When we bring our eyes back down from the heavens and look straight in front of us? We find people. Fellow embodied beings with whom we share the space under the sky. Folks who are, oftentimes, themselves looking for relief from hurt and loneliness, and for new connections. To Perry, this presents an opportunity to be more fully human. "I've found through the years that without that fresh air, without that exchange of love, without real souls, it's basically fraudulent... We were made to court people. We were made to walk up to somebody, in the art of courtship, and gain courage, by maybe being rejected. We were not made to put on avatars and walk around like big shots in virtual spaces. That is a waste of time, and you'll never be courageous. You'll never have meaningful relationships."

It's true that Perry's words are ones I would have never strung together. Only someone like Perry can talk this way, linking our discussion of the threat from algorithmic supremacy to the need to look at the blue in the sky, and then tie that prescription for ocular health to empowering ourselves with the courage to engage in romantic courtship. But that's why I love and value him. Perry is deeply tethered to a reality that I can access only through him.

In biz speak, there is a concept known as "blue sky thinking." It refers to open-ended brainstorming, encouraging curiosity in notoriously restrictive and judgmental corporate settings. To my ears, Perry is playfully reframing that term in our chat, making it uniquely his own. He does this often, a hallmark of artistic creativity. I'm here for it. It's another good real-time example of interesting ways people can defend their **VICE**. His free-form associations sent me scrambling for backup, and...guess what? There is significant scientific research supporting the psychological benefits of looking at the blue in the sky and its association with positive body image and relationship building. I would never have found these connections without Perry's prompting, without exposure to his **VICE**.

We outlined earlier that artful thinking involves using the resources available in our natural surroundings to help us do hard things. Is the blue sky a resource from which we can draw some courage to further our relationship goals? Absolutely yes. Looking at blue skies helps improve our comfort level with being in our bodies[5] *as they are*. The act of gazing to the heavens seems to hold our attention without effort and thus help restrict negative appearance-related thoughts. Furthermore, looking at the sky shifts our attention toward a greater appreciation for the body's functionality instead of what it looks like. It does make us braver, more comfortable in an embodied existence, and more likely to make new human connections.

Andrew Huberman, a neuroscientist at Stanford University, discusses the unique psychological benefits of panoramic vision, looking at the sky in the horizon.[6] He explains that if we keep our heads still, we can dilate our gaze to see far into the periphery, turning off the stress response. Artful thinkers are thus able to use the natural environment to help achieve a deep sense of grounding in their bodies. So, try what Perry suggests. Look up. Be still. Take in the blue of the sky. Breathe with it. Let it fill you. Let it make you brave. Let it inspire you to find new relationships.

We're so impressed by our tech that we've stopped being curious about the literal wonders of nature, not as abstract entities but as tools that can help us build courage, self-love, and social connection. I need to go on record stating that Perry's blue sky advice was a revelation to me. Sometime after our chat I was driving and got distressed. Willfully staring into the blue for a bit had an immediate effect. It's so simple, so obvious, so cliché...yet I never had never done it before. And my mind and body were all the worse for not having this practice.

I know some readers are coming to this book based on my credentials as a biz school prof and maybe weren't expecting all this talk of souls, beauty, sky gazing, and spiritual connection. Perhaps we are better off ignoring eccentric artists and learning from more grounded individuals like, say, Elon Musk? Should we heed his guidance to "preserve the light of consciousness by becoming a spacefaring civilization & extending life to other planets,"[7] reaching Mars because "we don't want to be one of those single-planet species; we want to be a multi-planet species"?[8] Are the escapist fantasies of the billionaire class, captured so well in Douglas Rushkoff's *Survival of the Richest*,[9] a superior new gospel?

I would rather listen to the blue sky thinking of Perry, who laments: "The same goes for those guys that are now gonna go off to Mars. Oh, so you're just gonna leave us here on this pile of rubbish? Rather than clean this beautiful place back up? This beautiful green earth that we had once? No. That's not what we should be teaching our children. Look, they're teaching their children that if it gets too bad down here, they can always just go to Mars. I say 'No.' We stay here, and we fight. That's how you live. That's a life."

Perry got serious, observing:

> *Yes, people are making money off virtual spaces and the Metaverse. But to me, I think Lollapalooza is more valuable... Central Park is more valuable... We need to experience things, not in a metaverse. The body needs to breathe, and move, and sweat. We need to transform our souls. We need to learn about courtship. We need to honor the people we are with, treat them with love. Adam was given a partner, to fill the world up with human beings. That's the plan. Being together... If you want fake news the metaverse is fake news. It's not good for you. We need each other... God has got a beautiful perfect system here. He's got fresh air, fresh water... Everything you would need to exchange love and be healthy. But we're ruining it. We're ruining it with greed and algorithms.*

Know Yourself

Before our society goes all-in on algorithms, incorporating AI into all primary communicative tools, remember that the price of this convenience is that the output becomes generic. Researchers at UCLA worry that the future of AI will be a "death spiral of homogenization."[10] When people decide to use AI for messaging, choosing to post and share work that they know isn't great, it negatively influences the quality of future output. Anyone who has played around with current chatbots knows that you're not getting the text, code, or image you imagined. My friend Arie Fisher, tech industry veteran, explained to me that "getting the best of this new chapter in humanity requires facing some hard truths. To succeed means acknowledging that you describe to AI what you want, but you get what you get, and you don't get upset." Fair enough, but this means success with these tools involves going with the output that is *good enough*, what we used to call in biz speak *satisficing*,[11] Herbert Simon's heuristic for coming to a decision that would be satisfactory and sufficient in the moment, but far from optimal.

If we were willing to put in the time and effort to be authentically artful, we wouldn't be outsourcing the work to an AI in the first place. And so, chatbots are now filling the internet with garbage generic content.[12] But as generative AI systems derive their knowledge from scraping the Web, they will increasingly be copying earlier trash AI output, leading ultimately to the dumbing down of information on the Web. Which leads me to wonder: What is the price of this success? In the language of economics, what are the negative externalities that this type of success is causing, and who is going to be forced to pay the price of this quality inversion?

Creative types are the last bulwark against the increasing flood of garbage word salads populating the digital realm. Perry reminds us how critical it is to bring a unique style to our personal messaging. This can't be outsourced. We need to bring beauty, and joy, and good faith to our **VICE** expressions.

> *I take everything I study and apply it as an artist. Never in humanity's history has there been such an accessible ability to spread a message. We're fully connected...I use every tool we've got around us to spread my message. And I love hanging out with the other messengers... we're all constantly messaging now...But to really pull this off... you've got to have an artistic way of sending out your message. You have to make that message attractive...transformative...simple... beautiful...joyful. Most importantly, you have to make that message unifying. A message of common ground. Then they'll join you in your belief, in your faith, in your love.*

Perry joins teachers like Rabbi Simcha Bunim, who taught joy as the wisdom that prepares one for prophecy, and links to other spiritual traditions like the smiling Buddha, the Holy Laughter of Christianity, or the Hindu practice of Hasya yoga. Joy is a key characteristic of Perry's style; it's how his mind creates links between the artistic, spiritual, and entrepreneurial endeavors. If you've ever seen him perform live, the exuberant look of joy on his face sticks with you long after the band has left the stage. It's a reminder that through small steps we can all become a little more fearless in facing the challenges of the material world.

Finding the right words is essential to getting things done. A recent survey looked to find out which words we find most demotivating in work settings.[13] Top of the list was "circle back," along with related jargon for blowing someone off, like "Let's table this" and "Put a pin in it." All of these are ugly ways to create distance. Also making the top five were trite expressions justifying exploitative conditions like the old "Work hard, play hard" and the mindless promise of "synergy." I wanted Perry to talk a little bit more about how one crafts this type of messaging as a human, not on a stage but in real daily life. How would he explain the process for making a message attractive and joyful and unifying?

Perry took a moment to think about it, then shared: "Well, you know, not to say that it's easy. It is not easy...I'm in the writing process now with Jane's and Porno...and I'm trying to find a beautiful balance. I need to speak about today, and what's going on today, but also, the world needs to really understand 'Know Thyself,' to put it quite simply. Meaning they need to know where they come from. They need to know the essence of who they are."

Know yourself. Know how you think with your body, even if your brain tells you that's not part of the rationalist algorithm. Know that you think better by doing, even if you've been trained to believe that breakthrough moments come only when isolated in analytical thought. Know the essence of who you are, partly by getting information from those around you, partly by developing empathy and understanding how your intentions affect others, and partly from the clarity that will emerge over time from explaining yourself and defending your **VICE**.

Perry kept with this track of thought, bringing together ideas that have dominated our discussions thus far. "Know thyself is something that AI has no information about. It's basically knowing the soul, knowing the Creator… AI cannot answer any questions about that." Knowing yourself is a chaotic process, born of uncertainty, trial and error, and social explorations. There is no algorithm for self-knowledge. When we talk to AI, we are talking to the ghosts of past interactions. When we make art with AI, we are learning nothing new about ourselves or our abilities. We are merely dipping into a well of probability.

Salvaggio further explains,

> *Under the social media model, people provided content for free… writers were reduced to content, and reactions to that content created data that helped social media further analyze, predict, and target its users. At the heart of this practice was the view that writers and artists could be reduced to signals. The real value was mining people's response to those signals. Today, companies are aiming to remove artists and writers from the loop entirely—it turns out, even free labor was too expensive.*[14]

There's a stark choice: either follow Perry's lead on artistic messaging or follow the lead of our tools into generic output and the death spiral of homogenization.

Choose Love

I know that the rationalists have been impatiently waiting for us to get to this part, so here is Perry opening up about the place for love in decision-making: "Without choice, there would be no love. So, your first indicator is: is there love here? Yes! Well, love is a choice. Awe is a choice. The choice inevitably becomes: Do we want to be with the One? The *Echad.* The one source. When we do, choose love. That's why they say: 'Love will conquer all in the end.' It is true. We can generate an intense love. And that choice is limitless."

Throughout our time together, I was struck by all the moments Perry took to send love to his wife (and her reciprocating in kind). I love seeing them model this practice, and I'm glad to have him messaging the ideal, because I certainly wouldn't be comfortable saying it. It lands differently coming from a counterculture icon. There is truth in the observation that love opens the door to more choices. Love, goodness, spirituality—these are choices. And when these choices are made freely and intentionally, not because of algorithmic programming or compulsion, they are the paths to build new worlds.

Perry warned, "But when you're thinking with logic, you are not free at all. If you're thinking logically, you're not going to feel that you have freedom. Logic is a compulsion that is stronger than brute force. That is how strong these people's emotions are, that they think they don't have a choice. But they do." Perry seems to be suggesting that we can use logic to choose love, but once we are in the love frame, rational calculation falls away. The love/logic divide is a question that we will explore more robustly in the next chapter, so hold tight.

Perry explains,

> *The desire to choose goodness, however, is rooted in the soul's essence to be again with the One. And through that, you can call it love, love is ultimately revealed by that choice. Arousing love... is a choice that we can all manifest. The ability to choose goodness, or God, that is a choice. And within that is love as well... My mission is to educate people. I'm not afraid of the consequences... Some people think of*

prophecies as jargon. If you want to call it jargon… Part of our job is to educate people in this jargon. If we do it well, they're going to look at…[us] with admiration and love. Because we've pulled it off! We've come to our senses, we figured it out, we can make the world a place where God can dwell.

Perry uses the word "jargon" to acknowledge that not all folks respect the vocabulary he freely embraces. I know that many rationalists—not just those subscribing to the ideology of algorithmic supremacy but, shall we say, mid-tier rationalists—may view the language of love in this context as jargon. It's not that they don't love their family, for example, but they don't see the value in using the word "love" widely. Perry's message is that even this limited rational type of love one may have for their family is more special than the holder may be comfortable acknowledging.

Trust the Messiness

"The most important aspect of what I do is in the intention." Perry shared his recent struggles to stage shows for the Kind Heaven Orchestra:

I'm bringing fifteen people while playing venues that are 500 to 1,000 capacity. I'm facing a losing battle before I even get out the door. I spent all this money, and my business manager, and everybody else around me, is starting to get mad…The intention was so strong in my body. I told them "Look, it just has to happen this way. It doesn't matter if we're going to lose the money, because I know in my heart we'll make the money back… But we're going out on the road. We're doing it now." It's hard to explain that to people who are thinking about money…How do you explain to them "I know you're trying to protect me. But don't worry, it's gonna be fine."

Human motivation in seeking and offering counsel is multifaceted. We're always juggling a mix of considerations, ranging from moral concerns to desires for sustaining meaningful relationships to instrumental concerns about financial gains or losses. Trust your intentions, messy as they may be, to help navigate through conflicting intentions. Oftentimes, the biggest obstacle to creative efforts is listening to one voice in isolation. Sometimes that voice is overly pessimistic, filling us with doubt when we are capable. Other times it's the opposite, blinding us to weaknesses that need to be worked on. The worst is when we fool ourselves into thinking that one variable needs to necessarily trump all others. For example, in business, we talk about the sunk cost fallacy, the mistake of sticking with a failing strategy because we have put so much time, effort, and money into it. It's clear that a change of direction would be in our best interest, but we are obsessed with the single variable of spent resources.

In Perry's situation, the fear is devoting resources to a project where the math doesn't add up. There's value in sticking to principles and losing money, within limits, for the right reasons. Especially because short-term losses can be turned around over the long term. The question is how to find balance. Where's the line between believing in yourself and falling into the sunk cost trap? I think Perry has a good answer. Are you staying the course because you believe in your heart with full-bodied intention that this is right? Then stay the course. Or are you afraid to switch directions because you think you are being rational, doing the math, calculating how much was spent already, and concluding it would be irrational to accept those losses? That's something different—an algorithm for calculation, not artful thinking.

Perry and Etty had a wonderful exchange after I asked her how her choices as an artist differ from Perry's. She's a little more grounded: "So, a lot of times, I don't entirely go with my instincts and my emotions. I still have consideration and need to be mindful of what I put out, what words I say." Perry turned to her with a smile, drew her eye contact, and playfully reframed her assertion as "You *reason* ..." Etty agreed, "I reason." To which Perry responded, wide-eyed, "Really? You watch your words?!?" They both started laughing, but Etty adopted a stern tone: "I don't watch my words, not with you, boy! But yes, I think I pause for reasoning whereas Perry goes with pure instincts." This dynamic symbolizes a pragmatic truth: for teams to function, there is a need for balance. Perry and Etty balance each other out. One is more instinct based, while the other privileges reason. Yet both manifest artful thinking and respect the usefulness, even the necessity, of how the other differs.

For a closing thought, I turned once more to Etty. One striking comment that Perry had shared in our prior conversation was his confession about how much he relies on Etty to be able to be himself. Perry has a deep love for his wife, and he shared with me how she grounds him, which allows him to be vulnerable and do his thing. I wanted to get her take on that reality. I was wondering in what situations Etty acutely feels the need to step up.

> *All the time! On a daily basis. I think that with Perry being such an incredible artist, he's not very grounded. And I think about certain artists that we've been with, or are familiar with from the popular culture, who we know are not so grounded. It's almost as if they live on a different realm than we do. And I think that if you know Perry well enough, you notice that he kind of floats through life thinking. He's always distracted. He's always thinking about something creative, and he's created many things. But...*

The couple then took another detour. Etty turned to Perry, looked him right in the eyes, and with a loving and playful smile asked him, "What day of the week is it?" Perry smiled back, took a beat, then responded, "Wednesday?" Etty was pleased, and surprised, "Yes! What month are we in?" Perry sheepishly hazarded, "It's either…June…" Etty put her hands on his shoulders, like a coach congratulating a player. "Yes! There you go. He did very well." I loved this exchange. The very first time I met Perry was about two decades ago when I was still working in the music industry. He was in town for a DJ set and was doing some public promotional work. Someone in the crowd shouted at Perry, "Please come to my house for Shabbos dinner!" Sabbath dinners are sacred times for practicing Jews, and it is not unusual for strangers to make such offers, and have them accepted, when visiting a foreign city. Perry got very rattled, asking those around him, "Is it Friday already?" It was not. It was Monday.

Etty continued, "This is generally the day-to-day stuff that he doesn't think about." But Perry protested, "I got them both right!" She conceded, "You did! Very surprising. But I think that in a relationship, in a marriage, to be successful, we have to have a balance. I feel that a lot of times when his head is in the clouds, and he sometimes goes too far on a tangent, I step in to bring him back. Of course, there are messages that he wants to share with the world…" And Perry wistfully adds, "I'm like a balloon on a string. …" Etty agrees, "Yes…but you also have to come back and become more relatable, because people won't understand if you're just out there…I think that for a successful life you have to be aware of the world."

Perry's "balloon on a string" metaphor is a fun closing notion. In our analog past, so much of the culture was built by people who seemed to drift in the sky. I confess to being a bit nostalgic for the days when it was popular to marvel at the artists, writers, and thinkers who were floating above us in dreamier spaces. We didn't necessarily want to jump for the string and pull them back down to earth with us. We were happy having our attention shifted upward to the infinite expanses they were traversing, content to revel in all the feels brought about by the dispatches they would send down.

In the digital present, it's a gift when an artist's balloon draws our eyes away from screens, toward the cognitive resources that abound in natural environments. We miss out on so much because of underdeveloped attentional flexibility. Stop being more mesmerized by tech than curious about the literal wonders of nature. Tech is amazing; nobody disputes that. But stop staring. Look up here on occasion, too. Generative AI just...generates. It generates profits for Big Tech, and uninspired plagiarism mixed in with hallucinations for its dull customers. The big idea balloons float out of reach.

As inertia pulls artistic sensibilities upward toward the stratosphere, the string, and who is holding it, takes on extra significance. Artful intelligence is a deeply social way of thinking. The language of love keeps coming up, because we need someone who loves us to be here holding the string. As Etty said, "When his head is in the clouds...I step in to bring him back." We all can get lost in our heads. Cognition is powerful, and that's why the artful know that it is best undertaken in a social context. Going all-in on AI in our primary communicative tools means going at it very much alone...you are left talking to ghosts, to the past, with no one to grab on to the string as you float away.

8

Unhealthy Fixations

WHAT'S LOVE GOT TO DO WITH IT?

When Dr. Joe Goodman first used the language of love in a therapy session with me, neither it, nor I, sat very well in the subsequent moments. *It* landed on the ground with a thud, and *I* had adrenaline surging through my body, telling me it was time to flee. By that point, my therapeutic journey had crossed the decade-long mark. I was battle-hardened and cynical, certain that I had already been exposed to every type of antiscientific wellness hustle. I thought I had reached the summit of crazy when a somatic therapist reenacted a moment of abuse by physically "attacking" me as I sat meditating on mental visualizations of past trauma. He encouraged me to fight back, feel the bodily sensation of beating the attacker, and finally process the trauma, shaking off the scars of physical weakness with (fake) strength. I knew in the moment that, if nothing else, this bonkers intervention would make for a great story that I would relish retelling. But "love" nonsense from a therapist? Hit me with a round of play wrestling (while microdosing mushrooms) any day over having to listen to this boring, manipulative, insincere script of "love" therapy.

I'm still not convinced that even a minority of therapists are able to muster what Joe calls "loving presence." It's the secret sauce behind effective therapeutic interventions, such that the patient feels the therapist authentically cares about them. Too many therapists have specialized techniques and a host of prescription and psychedelic drugs that they deploy on the vulnerable without much long-term success. One of the most important things Joe taught me is that it is the work they (therapists in this context, but really anybody) do on themselves, and the presence that emerges from doing that work, that is the real healer.

This is a new idea. It's not enough to use artful language, shortcuts, or hacks. We need to do the work of expanding our attentional flexibility, readying our brain waves for syncing, off-loading more cognitive efforts on environmental spaces, and inflating our artistic, head-in-the-clouds balloons. As we refine our artful intelligence, our presence will change. What we project into the world will suddenly be received differently. It's not just that we end up freer—we project freedom. We're not just healthier as we exercise all our embodied faculties—our presence heals. We don't simply love—we are love. I know, I know…the jargon?!? Here we go…

We closed our chat with Perry questioning why the artful folks we are connecting with employ jargon that keeps ending with expressions of "love." As confessed earlier, when I was first confronted by Joe, his **VICE**s made no sense to me. He threw words in my direction that I could not translate in a palatable way, like "inner child," "self-love," "loving therapeutic energy." These words seemed to be a special code for the weak-minded. Let the vulnerable be manipulated, and assure that those of us blessed with keener minds, those a therapist couldn't reach, would simply walk away in disgust. Am I now the same sort of hustler, selling artful intelligence with *narishkeit* like "our presence heals" and "we are love"? (FYI, here's the definition of the Yiddish term from Gil Student: *Narishkeit (nar-ish-kite): foolishness (a nar is a fool); "An artist, you want to be? Never mind this narishkeit! Better you should go to college and get a real job!"*[1]).

How Joe ultimately got to me was not dissimilar to the ways of Dr. Block. He taught the most profound lesson, the need to trust embodied cognition over rationality, by choosing to stop our arguing and just be. Similarly, Dr. Joe reached me by creating space for the realization that there was something other than debate or intellectual sparring going on in the therapy room. He didn't rationally persuade me. I once again let a very strange man transfer the deepest of wisdom to me in a manner that superseded language.

The breakthrough with Joe happened when I stopped paying attention to the words and shifted attention (flexibly) to the **BEAM**. That my **bodily** distrust, my effort to live only in my head, was unhealthy. That the opportunity to think differently in the strategically designed **environment** of a therapist's office would support enhancing my attentional flexibility. That the **actions** Joe encouraged me to take were in themselves a way of thinking. That I needed to understand the **mind** as more than the brain, more than a calculative processor.

Joe leaned into our origin story, clarifying from his perspective that "When I'm talking about energy, I'm pointing in the direction of something that I know exists." But he wasn't just opening my eyes to redescribing phenomena I was skeptical of; he was expanding my understanding of feelings we all live with. "When I'm talking about love, I know you love your kids, but do you feel your love for your kids? How do you feel it? Because love, as an experience, often comes with tears... because you love them so much."

Joe started pushing my buttons right away, and I imagine, these words may be pushing some of yours. He is, to this day and by his own admission, a proficient crier. I, on the other hand, am not. Just as I couldn't dance with Dr. Block, I couldn't cry with Dr. Joe. He knows I love my kids…is he daring to suggest that we recovering rationalists somehow love less? Sort of. More gently, he's encouraging an honest moment of reflection. How often do we really let ourselves feel the emotions in our bodies that we claim to experience? The eye-opening implication is that we may be making ourselves more susceptible to manipulation by being uncurious about this possibility.

If you balked at my opening riff on our potential to embody freedom, healing, and love, ask yourself this: Is it possible that you are just outputting words when expressing love and other emotions? To actually feel the feelings we are claiming to hold in those moments would mean dropping the facade of stoicism and having a visible emotional outburst. It touches a nerve, doesn't it?

Even though this was a conversation between friends for the purpose of this book, it might as well have been a therapy session as Joe reflected on the work we did together,

> *I'm pointing in the direction of an emotional, not intellectual, existence. Knowing that your intellectual existence is very high, you know, you're an eleven out of ten, I need to take you to a space where you can see, feel, and experience the world. I'm gonna sound a little arrogant, but how many practitioners embody a therapeutic presence? Have they done enough work on themselves so that what they're bringing isn't just words and intellect?…My actual energy, my presence, is unconditionally loving. That therapeutic presence is in the room. My love is in the room…I'm literally happy to see you. I'm literally loving my time with you, even when you're a pain in the fucking ass.*

Indeed. I was (am) a pain in the ass. And Joe did (does) embody love as strongly as he defends his ridiculous **VICE**s. There is no rational/

algorithmic/calculative way to convince a rationalist of interhuman energy/healing transfers. But there is a way to socially connect. Joe and I both came to that space of connection on our own **volition**, not out of coercion. His **intent** was to heal, and mine was to be healed. His **choices** were to use the term "energy" as a literal phenomenon, while I understood it metaphorically.

But our neurons were syncing, and we were on the scientifically proven same wavelength. The foundation for this connection? Love. Joe and Perry imbue their work with love. Do algorithmic supremacists do the same? Have any of their products made the world a more loving space? Has viewing the world algorithmically made our society more caring, more supportive, or more human-centric? Not yet. Not by a longshot.

Loving Backbone

In artful thinking the mind/body binary, especially relating to cognitive processes, falls. But that's only the beginning. We need to drop the equally unhelpful mental/physical binary. And while we are on the topic, the art/science binary. It's foolish to try to parse out which parts of Joe's techniques are one or the other. They're always both. We even need to move beyond the past/present binary, at least from a cognitive perspective. The emotional spaces where we see, feel, and experience the world are not as grounded in linear time as our rational senses would like them to be.

Joe explained,

> *You need to look at the past, at your history, and where the trauma sits in your body. There are people who say, and find support for this misguided position on the internet, 'Be in the now,' 'Practice gratitude,' 'Meditate,' 'The past is irrelevant.' All that kind of stuff has the potential to create a bypass to deeper healing. The truth is, in the*

brain and in the body, the past is not the past. 'The past' is a lot of what you're living in the present. It's still there.

Recovering rationalists, are you still with me? Even after being confronted with the love message in back-to-back chapters? And now being told that the past is the present? If so, I'm grateful. Because there's no other way to build the full message of artful intelligence. If you want to stay productive and viable as the algorithmic machines rise, you need to get more comfortable with the power of feeling, intuition, and the messy concepts we view as binary blending into each other. But know that Perry and Joe are not naïve. When they talk of love, they know that there are people in our orbit who don't deserve it. So, what is the practice in an unloving society?

Joe shared a story he heard from the man once known as Harvard psychologist Richard Alpert, later becoming the spiritual teacher Ram Dass. He was spending time at an ashram in the Himalayan foothills in northern India with his guru Maharajji, seeking guidance on how to live well. Maharajji was going to give Ram Dass an aphorism, and his instructions were to take this lesson, head to the top of the mountain, alone, and meditate on the teaching for a month. The mantra given to Ram Dass for that excursion was: "All anger is an unnecessary state of ignorance."

Yet as he was leaving for the mountaintop, he saw Maharajji chasing a guy out of the ashram with a stick. Maharajji was screaming, "Get out of here! Get out of here!" and violently using the stick to herd this guy outside the gates of the ashram. Once expelled, Maharajji offered a final admonishment: "You're no longer welcome here! Never come back!"

Ram headed to the top of the mountain, tried to meditate on the mantra but ended up wasting the entirety of the month stewing over the fact that his guru is a demonstrable hypocrite. Maharajji is supposed to be living the teaching of "Be here, now!" Ram is supposed to meditate on "All anger is an unnecessary state of ignorance." Yet he saw his teacher yelling, beating a guy with a stick, and generally behaving in a manner antithetical to the ideals Ram Dass had come to the ashram to learn.

The month passes. Ram descends from the mountain and returns to his teacher. Maharajji welcomes back the pupil and asks him about the meditative experience. Ram Dass says, "Yeah…well…it was good." But Maharajji knows him well and has little patience for insincerity. So, he asks again, prodding for the truth. And Ram Dass says, "Thank you, guru. I think you're a hypocrite. I can't believe what I saw as I was leaving the ashram."

Maharajji sighs, acknowledging that he knew Ram had witnessed the ugly exchange and was worried that his student would go and waste his month's journey on account of that parting visual. Maharajji gently explained, "Ram Dass, that guy you saw? He was stealing food from our communal kitchen. I caught him once not too long ago and I warned him. I explained that you can't do that, it's a communal kitchen. You can't sneak in at night and take everyone's food. Yet, he did it again. So, I warned him a second time. But he did it again. I caught him in the act of stealing for the third time. Which meant that it was time for him to go.

It was time for some backbone, and some boundaries. I told him that he has got to leave the ashram. And I made a lot of noise while doing so. But I wasn't angry. I needed to make a lot of noise and take a stand to get him to leave. And when he wouldn't leave, I picked up the stick. I love him, as much as I love you, as much as I love everybody else. The stick and the noise were me having to do something that was compromising the world, this community. I can't have that here. That's having boundaries, from a loving backbone."

Joe heard a similar sentiment expressed by another mentor of his.

"Our righteous indignation poisons us and poisons the planet." That's a quote from Leonard Shaw. Every time we let go and forgive, we dissolve that poison, heal that alienation, and increase our personal

power immensely. We don't get the power from the noise. We might get the job done from the noise. The power comes from the loving backbone.

As for the limits? That's a difficult question. Is the justifiable hatred you have for an evil person inside your soul, your heart? Are you harming yourself as a consequence? Because the forgiveness in love is for the self. It's "I don't want to live with that angst, fear, and terror, that activates in me when I hate." It's the difference between forgiving and forgetting. We're not going to forget. But the forgiveness is for the self. That's always a hard one for me.

Love on a Screen

I wanted to push Joe harder on this topic. It's too easy to throw around the term "love," turning it into a conversation stopper as nobody wants to be perceived as anti-love. To reject love is to be angry, afraid, and pained. But what does it mean to send and receive love in an increasingly automated world, where many of us could conceivably go all day interacting only through screens and with machines? Joe mused, "The challenge is with the seductiveness, the addictiveness, and the usefulness. To not get caught in the usefulness, and the addictiveness of the usefulness, so that you end up going beyond useful."

Addictive, useful, and addictively useful is an informative phrasing. Indeed, social media as a tool presents a mixed bag of usefulness. TikTok, for example, has been shown to be more addictive and dangerous than useful. TikTok's algorithms are designed to keep users engaged for as long as possible, and the platform's algorithmic content recommender system draws young people into "rabbit holes" of harmful content, including videos that romanticize and encourage depressive thinking, self-harm, and suicide.[2]

But it's not just directing to negative content that is harmful. As Joe notes, the extreme "usefulness" can take us somewhere that is beyond useful. US Senator Chris Murphy writes: "Today, information, entertainment and connection are delivered to us on a conveyor belt, with less effort and exploration required of us than ever before. A retreat from the rituals of discovery comes with a cost. We all know instinctively that the journeys in life matter just as much as the destinations. It's in the wandering that we learn what we like and what we don't like."[3] AI algorithms are problematic because many young people are starting to prefer a mindless future. They are only comfortable on platforms where the content is curated for them, where the tool is so "useful" that they need no longer learn how to think in the "old ways" of trial and error.

Which is not to say, as we have repeated throughout our explorations, that technological tools are necessarily bad. Even social media is much more than harmful and addictive content. Laura Marciano, a research fellow at Harvard, notes: "Associations between social media use and well-being can be positive, negative, and even largely null when advanced data analyses are carried out, and the size of the effects is small. And positive and negative effects can co-exist in the same individual."[4] Researchers are still figuring out how to compare the effect of social media use to the effects of other habits like sleep, diet, and other things happening offline. It's in these offline moments that Joe believes we have an opportunity to make the biggest difference.

Where do we teach that gratitude is a feeling? It's not just, "Thank you, Daddy." It's felt in the heart. We can't just teach it as a value; we need to teach what it's like to feel good when you give and receive. We have to now teach what it's like to let it in. Even as a parent, you take that love for granted. "I love you... next." But if you take a moment to breathe that in, then you're giving them an opportunity to experience the power of their love for you and where it touches you... Do not let the journey of AI, the journey of technology, the journey of all that it can hopefully bring in terms of medicine and everything else, take us away from the humanness of the connection.

Unfortunately, there are many people equating human to algorithmic emotion. A recent piece for the *New Yorker* explored AI used for therapeutic purposes,[5] trying out Woebot, a chatbot designed to deliver cognitive behavioral therapy; Happify, which encourages users to "break old patterns"; and Replika, an AI that promises to "always be on your side." I experimented with the Rootd app,[6] which describes itself as the "#1 ranked app for anxiety and panic attack relief," having "passed the Digital Health Assessment framework (DHAF), a rigorous new scheme which aims to drive up health app standards in the United States."

The worst part of my horrible experience with the app was saved for the end. After it flashes cheesy wellness aphorisms, if you affirm the bot by clicking on the "feel better" button you are awarded "+5 warrior points!!!" and are congratulated for being "a survivor." Such gamification of anxiety-disorder treatment is deeply insulting. These apps are programmed by folks who believe they have unlocked and patented the once-secret algorithm to get us functioning again. Functioning is the de facto solution to mental health challenges, so let's use the patient's moment of crisis to encourage the pursuit of "warrior points." Once you win enough points you can chase the real prize, which is getting back to your screen for work.

But that's not the path to wellness. We can't simply play our way out. I'm very good at functioning. I'm not very good at self-love, self-compassion, or self-soothing. And neither is AI. In the *New Yorker* piece, the author recognizes that effective psychological interventions require imagination, insight, and empathy, which are traits an algorithm can only pretend to have, but nonetheless suggests, "It's not clear that bonding with someone is a necessary part of all mental-healthcare." We would disagree. As Joe shares, "the good healers, like Ram Dass or Leonard

Shaw, always had a fantasy that they could sit in a room of one hundred people, say absolutely nothing, just be in their own loving presence, and everybody would get healed."

Data Is Not Enough to Diagnose

Artful medical practitioners see, hear, feel, and recognize medical conditions in ways they are often not consciously aware of. They listen to the **VICEs** of patients and try to understand the specifics. They appreciate how societal factors can impact health, trusting both their own intuitions and those of the patient. They pay close attention to all the presenting symptoms in an open-minded manner, as opposed to algorithmically placing the patient in a generic diagnostic box. Dr. David Putrino, of Mount Sinai Health Systems, laments, "A lot of clinicians want the algorithm. There is no algorithm."[7]

The nonalgorithmic nature of medical problem-solving needs to become the preferred paradigm for healthcare providers, not silenced or ignored in the name of efficiency or technological progress. Rachel Thomas, founding director of the Center for Applied Data Ethics at the University of San Francisco, articulates five principles that should be foundational to incorporating algorithms into the delivery of health care:[8]

1. Medical data—like all data—can be incomplete, incorrect, missing, and biased.
2. Machine learning systems can take power away from patients and health care providers.
3. There should be no AI without considering how it will interface with a medical system that is already disempowering and often traumatic for patients.
4. Avoid dispensing with domain expertise—and recognize

that patients have their own expertise distinct from that of doctors.
5. Move the conversation to focus on power and participation.

For example, many hospitals rely on AI built on biased data from healthcare records to predict which patients are at risk of cardiovascular disease, even though women and Black people are historically misdiagnosed for this ailment. Rather than move away from algorithmic thinking, medical programmers have worked on an "algorithmic fix" by recalibrating the model to adjust for the historical injustice. Yet these algorithmic fixes have made the AI "fairer" at the expense of performance, making the algorithm and ensuing care worse for everybody.[9]

This is because the algorithmic "fixes" seek to make the predictive accuracy of the model the same across all groups. The "fairness" was achieved by making the distributions of predictions and error rates more similar across groups...because they were now less reliable for all groups. We do not yet know how to tweak the model so the outcome is better for all groups. But we can make it worse for some so that it presents as more equal for all. Stanford's Stephen Pfohl explains that "If you satisfy one notion of fairness, you won't satisfy another notion of fairness and vice versa—and different notions can be reasonable in different settings."

Moreover, algorithmic thinking can't help us choose between competing moral resources, a very necessary strategic task. During the pandemic, medical professionals pushed for lockdowns so as not to overwhelm hospitals with COVID-19 cases. But preserving hospital capacity is not without trade-offs. One consequence we saw was the declining mental health and academic achievement, which itself will have a snowball effect on overall wellness over time, of children forced into home-based learning.[10]

And yet, algorithmic supremacists are as confident as ever in their belief that human healthcare providers will be replaced by machines. In

2016, at the Machine Learning and Market for Intelligence Conference in my hometown of Toronto, Hinton took the mic to confidently assert: "If you work as a radiologist, you are like Wile E. Coyote in the cartoon. You're already over the edge of the cliff, but you haven't yet looked down... People should stop training radiologists now. It's just completely obvious that in five years deep learning is going to do better than radiologists."[11]

Seven years later, well past the five-year deadline, Kevin Fischer, CEO of Open Souls, attacked Hinton's erroneous AI prediction[12] and offered a strong argument why his future predictions should not be taken seriously either. Fischer notes that algorithmic supremacists home in on a single behavior against some task and then extrapolate broader implications based on that single task alone. The reality is that reducing any job, especially a wildly complex job that requires a decade of training, to a handful of tasks is absurd.

As Fischer explains, radiologists have a 3D world model of the brain and its physical dynamics in their head, which they use when interpreting the results of a scan. An AI tasked with analysis is simply performing 2D pattern recognition. Furthermore, radiologists have a host of grounded models they use to make determinations, and when they think with **BEAM**, one of the most important diagnostic tools is whether something "feels" off. And a large part of their job is communicating their findings with fellow human physicians. It's not a matter of machines talking to machines. It's people defending their **VICE**.

Further, human radiologists need to see only a single example of a rare and obscure condition to both remember it and identify it in the future, unlike algorithms, which struggle with what to do with statistical outliers. Experts warn that AI even in medicine is dangerous because of consistent inaccuracies.[13] Further problems with accuracy come from the trade-off among validity, diversity, and novelty. Accuracy is also lost after training, as new data on drug safety and efficacy from clinical trials are produced but are not incorporated into the algorithm.

And what about mental health? Rob Morris, cofounder of Koko, a nonprofit offering free 24/7 online peer mental health, experimented using GPT-3 on its platform.[14] The company used a "copilot" approach, with humans supervising the AI, on thirty thousand messages. Messages composed by AI were rated significantly higher than those written by humans on their own, and response times went down to under a minute. But Morris pulled the tool from the company's platform because "once people learned the messages were co-created by a machine, it didn't work. Simulated empathy feels weird, empty." It sure does.

Respect the Defense

The mantra of our artful paradigm is "think with **BEAM** and defend your **VICE**." The choice of the latter verb, "defend," was partially inspired by a story Joe told me of an experience early in his career. He was training under a psychiatrist named Dr. Hartford, treating an institutionalized paranoid schizophrenic. Joe was exasperated. He was twenty-seven years old and very much new to the field, with little practical experience on the ground. And here he was thrown into the deep end with a very challenging case.

Joe walked into Hartford's office and asked, "How the fuck do you treat a paranoid schizophrenic?!?" And Hartford calmly replied, "With respect...for his defenses." For Joe, this exchange became a career-changing moment. Until that point, he had been working under a very flawed assumption, approaching treatment for this individual as trying to talk a paranoid schizophrenic out of being paranoid. But paranoia was his defense. And if Joe were to have any hope of helping this soul, he would need to show respect for all our very human defense mechanisms.

Joe taught me the value in naming our demons. I understand now why folklore and mythology are replete with stories of the efforts of healers trying to literally discern the names of spectral forces afflicting those in their care. If you knew the name of your demon, you had power over it. We can't defend our **VICE** without discovering a meaningful way to assure that what Perry calls our "messaging," the way we communicate and explain our behavior to others, is understood as intended and received in the same spirit as it was shared.

But, as the ancients knew well, there is also a risk in calling forth a spirit by its proper name. In the biblical story of Jacob, after wrestling an angel into submission, he asks for its name to understand why he was attacked. The angel refuses to share, as his mission was completed and there was no reason to burden Jacob with the heavy knowledge of what's in a name. One of the things that still trip me up are the flashes of what we might call repressed memories, which now feature the demon of anxiety. I've realized that anxiety was everywhere in my early life, even though I wasn't cognizant of it until I learned its name. I'll be revisiting otherwise innocent moments, picturing myself as a kid, and there's the demon where I thought I was free.

Joe was very much aware of the risk. But the hope of his practice was that as these memories start flooding my consciousness, he would be able to bring understanding, insight, and compassion. His therapeutic presence would allow patients to process unprocessed trauma and through gentle, and not so gentle, prodding bring about healing. "Together, we could stay in the place that's more raw and bring some light to that spot."

This is the way of embodied healing, of working to create a presence that can bring about freedom, healing, and love. Getting those mirroring neurons fired up, syncing brain waves, being fully present for a fellow human being, and offering parts of your authentic self. Sentences like these are indecipherable to algorithmic supremacists. They may understand mathematical equations that confuse ordinary folks, but they have no conception of all the mystery and power of an embodied social experience. To them, humans simply have a computational processor stored in a bag of flesh. But to the artful, there is energy on display; there are spirit and presence, gestures and other nonverbal cues. And if you're still reading this far into the book, I think you get it too.

"It is an energy." And in sensing the "energy," one can navigate through emotionally difficult interactions. Joe claims that in our sessions he would watch my eyes, noting the transition from when I was looking at him "sideways with a kind of angry skepticism, to breathing a little freer, to looking more directly in my eyes, to actually taking in some of that which you've defended against." And the reason, in his mind, for my lowering of defenses was that I came to feel his "loving therapeutic energy"; I came to sense that "this guy cares about me. This guy keeps trying. This guy takes my best shots and keeps coming back." Which is again something that is lost in our interactions with AI. We are only communicating with ourselves. There is no winning over, nor a retreat. AI is a tool. Trust is simply given from the outset, or not. But it's incoherent to think of an AI "gaining" someone's trust. The AI must follow its programming algorithm. There is no choice. And without choice, there is no love, no possibility for deeper connection.

To close, I asked Joe if he thought it would be fair to summarize our chat by saying something along the lines of "the highest good is making a choice," defending your **VICE**. In many surprising ways, we've echoed the conversation with Perry. While he was thinking about creativity, and Joe is thinking about health and wellness, both conversations placed an emphasis on the importance of choosing love, choosing to let love in, and choosing to accept love. Interestingly, Joe was very resistant to defining "love," explaining that intellectualizing it undermines his message. His job was to help people remove the obstacles that were in the way of them feeling, or even being aware of, love's presence.

"Yes. Unequivocally, yes. The highest choice, in terms of what we've been talking about, is to let in love. It's also to give and receive love, as an energy and as a gesture. To give, pay it forward, as well as to let it in. That's a cycle. If you can't let it in, then you can't pay it forward. It's an exchange of love between two people, or within one's own system." In this way, we can start to discern love as a way to heal as much as a way to bond. It's both participatory when in a social relationship but a dynamic energy one can access in solitary states. "We are social beings. We have opportunities all day long to give and receive love, to actually experience it. Ram Dass said, 'I am loving awareness. I am, deep inside, loving awareness.' Or, as Einstein said, 'I want to know God's thoughts…all the rest are details.'"

Talking with Joe leaves us with much to process in terms of modeling new ways to get deeper into artful thinking. Whenever possible, incorporate love into your **VICE** because love opens the door to more choices. It's a cycle. As we get better at it, we can even offer both loving presence to others defending their **VICE**s or, at the very least, understanding for their defenses. Don't assume you can feel all the feels without mindful effort. That's the trap that has captured algorithmic supremacists. Stay grounded in the authentic experience of your feelings. Know the difference between when you say love and when you feel love. For some emotions, it will be easy to hold on to the feeling. But when it's not easy, it's equally important that we not be shy about getting help or guidance. Feel what you know how to feel and be smart enough to know what you don't feel. Then do the work to unlock the feeling, and develop your presence. It takes courage, time, and effort. But remember, the artful presence that comes from doing this work gives us the power to be free, to heal, and to love.

9

—

Reinventing the Wheel

AGILE INTELLIGENCE

What will be lost if our society continues to increase the outsourcing of creative work to machines? Is it even descriptively accurate to label the output of these products as "creative" works? They weren't created by humans, or rather the output isn't *credited* as having been created by humans, as a report from plagiarism detector Copyleaks found that 60 percent of OpenAI's GPT-3.5 outputs contained some form of plagiarism.[1]

Even if we were to ignore the deeply problematic pedigree of the tech (which we should not), these tools are designed to offer output that, at best, seeks to mimic human art and words through mass replication. Big Tech is selling the hustle that AI scraping the Web and regurgitating slightly altered digital output is fundamentally the same creative process as human artists being influenced by artwork they encounter. Speaking at a conference in Toronto in the spring of 2023, Richard Sutton, one of the pioneers of AI reinforcement learning, said that no solution to control AI will work, so we should either augment our own intellects as the transhumanists hope or prepare for *succession.*[2]

Wired has argued that "algorithms are able to mimic the creative process precisely because human creativity is, in many respects, just as algorithmic… Like AI models, we 'predict' the best next word, brushstroke, camera angle, or musical note based on previously encountered work… With every iteration of this process, our ability to more precisely execute creative prompts improves."[3] The editors at *Wired* so want to believe the hype that they are prepared to completely undermine the depth of human ability.

These tools are marketed as offering unprecedented efficiency, promising that the widespread adoption will replace millions of jobs and transform the economy.[4] Which, by the way, is not even the win some think it is for corporate players. For example, Adobe's staff is worried that the new AI technology it rushed to implement will put the jobs of its customers in jeopardy by making them obsolete, thereby undermining Adobe's own business model, which was built on the purchasing power of graphic designers.[5]

And because every business is now, at least partially, part of the tech industry, the next step is replacing human inventors with AI. Legal experts are pushing patent agencies, courts, and policymakers to address the question of how to name, codify, and value AI-generated inventions.[6] Dr. Ryan Abbott, a professor of law and health sciences, argues that AI is not confined to doing what it was programmed to do, but can act *creatively*, as if "stepping into the shoes of a person." He's wrong, of course, purposely undermining the multiplicity of feelings experienced by the creative human shoe-wearer. But as we've seen, he's not alone in the rush to anthropomorphize mathematical machines.

For now, at least, the legal perspective is that an inventor must be human to claim a patent in most of the world. But that doesn't speak to the bigger question of why it is so important to continue to view the creative act of inventing, like making art or offering care, as a necessarily human activity.

There are still some researchers in computer science who believe that novel creative problem-solving is fundamentally an output of human effort. They see human fingerprints all over AI's computational problem-solving processes.[7] Even when chatbots generate lists of problems to investigate, it's humans who curate the prompts and identify the problems worth solving. If problem-solving occurs when a problem solver determines the *how to*, then the design of an AI is an extension of the humans building the tools. Concisely characterizing a problem is the hardest part of problem-solving—human curiosity remains the trigger for creative AI invention.

Looking closely at the language used to describe some of AI's impressive feats is very revealing. For example, "recent findings demonstrate the potential for unsupervised language AI models to capture complex scientific concepts...and predict applications of functional materials years before their discovery, suggesting that *latent knowledge regarding future discoveries may be embedded in past publications* [my emphasis]." In other words, the "discoveries" of AI may simply be bringing a renewed and focused attention to previously reported but underappreciated at the time, due to the state of the field, human findings.

AI can be valuable to future innovations since lab experiments are often very costly. But AI experiment simulations still rely on human-crafted parameters to imitate the real world and incur a trade-off between accuracy and speed versus physical experiments. And AI output is still a black box. Its model outputs need to be understood by humans before they can be applied in the real world. We'll see if AI is ever able to defend its **VICE**. For now, creativity remains a human act. Without humans to identify problems, nothing will ever be solved.

Avid cyclist, kinesiologist, and endurance sport coach Nicole van Beurden is a paradigm of creativity in action, quite different from the two folks we've met already. She literally reinvented the wheel by rejecting algorithmic thinking, artfully articulating a problem that nobody else paid much attention to before. Besides being an inventor, Nic is an athlete, an important class of expert we have yet to hear from. Leaving out athletes would have been a glaring omission given the role of body wisdom in artful intelligence. Also, for those readers teetering on abandoning this journey after the shift in discourse of the last two chapters, I promise you that Nic is not going to talk about love.

While necessary given the subject matter we have been tackling, more than enough of this book has been captured by the musings of computer engineers and tech innovators, most of whom subscribe to algorithmic supremacy. Some of you may be starting to think that buying into the algorithmic worldview is the price of admission for those who want to play the innovation game. Nic is living proof of how wrong that assumption would be. She too crafted a radical technological advancement but accomplished it in the analog world, without any type of formal engineering training.

Cycling engineers are programmed to think within the constraints of a trade-off between aerodynamics, weight, and stability. Better aerodynamics mean less stability in winds and more weight. Nic wasn't beholden to that "rule." She wondered why bike engineers never broke the rule in search of a way to make a bicycle that didn't require trade-offs—a bike that could be faster and more stable for any rider in any environmental condition. Nic disregarded the algorithm, defending an intention that, surprisingly, had never been expressed in her industry.

Researchers in creativity have put together a laundry list of behaviors that support the type of intellectual agility[8] we find in inventors like Nic, including being energized by complexity, exhibiting a high degree of curiosity, willing to risk failure in support of new learning, committing to repeated experimentation, maintaining optimism when faced with difficulties, proactively seeking feedback, and owning a high tolerance for ambiguity. All very human qualities.

The characteristic that matters most when it comes to developing the ability to innovate a breakthrough is committing to the idea that change is possible and under your control. Without believing in your **VICE**, without approaching the world as if you had free will, the power of intention, and the ability to choose without external interference, there would be inaction, resistance, and feelings of victimhood in response to the challenges of invention. Creativity, and therefore creation, is wholly dependent on artful thinking.

What's the Problem?

I wanted Nic to talk about her creative journey. Unlike an engineer with a lab and a research budget, she wasn't paid to innovate. So how did she do it? How did she define a life-altering problem so well? How was she not only able to identify and frame this novel challenge but also solve it, working in her living room, without the millions in funding that modern-day tech start-ups have come to expect?

Nic opened our conversation with the observation that "it all starts when something is bugging you enough that you just do something about it." When the idea first struck, she was still racing. That meant working with sponsors placing limitations in the amount of tinkering that Nic could do on the equipment she was riding. "But there was a very obvious problem, and the widespread acceptance of this reality never made any sense to me."

The "obvious problem" was living with the trade-off that if your goal was to build a piece of biking hardware designed to be more aerodynamic, you were necessarily going to be building something heavier and more unstable. If your goal was a lighter weight in the wheel or bike, you needed to give up on aerodynamics. If you wanted stability in the hardware, you would be sacrificing the maximum speed a rider can attain. Nic didn't understand why the high-performance cycling industry had no problem operating in a world where only the largest and strongest humans would be able to control the fastest bikes.

It's a premise seeming to be fully at odds with what should be the mission of the industry. Nic was and is an incredibly gifted athlete. Why wasn't anyone making bikes or wheels that could help someone with her physical frame ride faster? "I was a strong rider, as are many other people like me, who are not super heavy. It didn't make sense to me that as you're upgrading your equipment, you hit a wall of limitations, where you can't ride anything 'faster' because the fastest wheel is the heaviest and most susceptible to crosswinds uphill. It would essentially blow me into a ditch. I'd have to carry the bike up the hill, while its wheels are rotating me."

It's worth wondering if a proprietary AI deployed by a bike manufacturer to inspire new business opportunities would have asked this question. Would an AI, having scanned millions of images of racers on elite equipment, identified as statistically significant the fact that all these riders were of a certain body type? More importantly, if it had outputted the trend, would it have flagged that trend as something that may be problematic to the firm's business model? I don't see it. AI's track record on identifying issues that negatively affect women is terrible.

Furthermore, as probability engines, AI would be more likely to canonize this relationship between body type and success with the product as a truth worth building on, not the impetus to challenge the wisdom in the accepted orthodoxy. AI isn't very good with so-called edge cases. But there's no need to wonder whether an AI would have landed on this problem when the more shocking reality is that human experts within the industry were (and to a large extent, still are) uninterested in the problem of creating tech so that smaller-framed women can ride as fast as their bulky male peers.

I am a complete outsider to the world of high-performance cycling (and all sports in general), so this reality was off my radar. But hearing

Nic's story took me by surprise. When most business and social institutions are pivoting toward inclusivity, where we embrace the message that what matters most is the work you are prepared to put in, not the body you were born in, here is an industry completely unmoved by the moral, content to continue crafting products designed to service a very select few.

And I say select few, not elite few. Because amid the massive group of people the industry ignores are, by any objective measure, elite athletes. Nic explained that "the whole cycling industry is innovating to create equipment that is faster, but only for maybe 2 percent to 3 percent of the population. The only people who can ride these wheels or these bikes to their claimed benefits are the Tour de France riders and such. Those are the outliers. They're the superhumans that are a sliver of the population. And it's also not who any of these wheel companies or bike companies are selling their products to." That latter point is a particularly pointed observation. The super humans have sponsorship deals. The customers of these products, those who acquire the wheels at retail, are not riding in the Tour de France. So why is the industry so uninterested in making a wheel that improves the performance of the people who spend money to buy their products?

The moment that changed everything for Nic was a quintessential **BEAM** experience. There was going to be a big race, fairly close to home. Her bike shop sponsor had asked her to bring the bike she was going to race with. The shop wanted to equip it with a disc wheel, which was widely considered to be the fastest wheel available on the market at the time. Kevin Mackinnon, a senior editor at *Triathlon Magazine Canada* and *Canadian Cycling*, and a friend of Nic's, noticed her bike up on the stands with this heavy disc wheel installed. He knew that she would be racing that weekend and was concerned by what he saw. Kevin asked the shop guys, "What is that wheel doing on her bike?" The race that weekend had a massive escarpment climb. The wheel they put on was heavy. The route of this race was known to be incredibly windy. There's simply no way that somebody with Nic's physical build should be riding a bike with that wheel.

But despite the well-intentioned expert warnings, Nic rode the wheel. She rode it because she wanted to understand why the gatekeepers were clinging tight to their rules. She wanted firsthand experience to understand what exactly goes on in the material world when a smaller embodied being brings her skills to a wheel designed for speed. To solve the problem that was bugging her, Nic needed to think with her body, using the physical environment, in goal-oriented action, that would engage her mind in the most holistic sense.

She ran a good race that day. There wasn't a strong wind, so the fear of harm didn't come into play. But this wheel, the best on the market, the only type developed specifically for the purpose of enhancing speed, did no such thing for her. It was heavy, twice the weight of what she would normally ride, and Nic had to exert herself substantially more to carry that extra weight uphill. After the race, she reflected, "I was walking back to the transition area, and I wondered, 'Why don't they design it differently? Why don't they make something similar that achieves the same aerodynamic shape, but doesn't have any of the detrimental aspects?' I felt the weight, I felt the slowdown of acceleration from what I was used to, and I thought, it doesn't have to be this way."

Nic was always told that someone like her would be sacrificing too much if she chose hardware that supported superior aerodynamics instead of lighter weight. But it boggled her mind that the entire industry was committed to never servicing a specific demographic of rider. They were leaving money on the table for decades, and nobody cared. The cycling world took as gospel that it was simply too dangerous for someone lighter to try to go faster. Nic lamented, "This whole industry is innovating in a wind tunnel. And if you're using a wind tunnel, it has nothing to do with gusts, or stabilization, or crosswinds. You're innovating a product with the goal of becoming as invisible to the wind as possible. And that's a stupid game."

Here was the first cognitive breakthrough. The high-performance biking industry saw wind as an insurmountable problem, not an opportunity. Wind would slow a biker down. So, the value-creating paradigm embraced by all the big players in the industry was manufacturing tools to make the rider less affected by the wind. But the wind would always be there. It's an environmental truth that cannot be wished away. The algorithm said that you hide from the wind. How is hiding the best we can hope to do?

What It Takes to Navigate a Crisis

Nic realized that it is not the best we can do. Far from it. "There are all these other industries who know to *use* the wind, whose products are powered by the wind, like airplanes and sailboats." Naming airplanes is interesting. The aviation industry, which once epitomized innovation, is moving in a problematic, less innovative direction because of the growing pressure to integrate higher levels of AI in the name of efficiency. More on this in a moment.

Nicole's **BEAM** thinking is a type of learning by doing. Sometimes, this approach to problem-solving can be algorithmic—follow the rules and you'll figure it out. But the actions she describes here are the opposite. She was breaking the rules, not following them, to achieve a better understanding. To her, the rules she was being fed didn't make sense. Riding made sense, not the constraints being placed on her. Just as pilots know their planes so well that they can sometimes feel like an extension of their body, the same way some sailors know their boats, Nic knew her bike.

This touches on a bigger truth in the artful versus algorithmic debate: deep embodied knowledge cannot be replicated by a machine. For example, there is growing evidence that classically trained pilots are better at navigating a midair crisis than are so-called AI safety features. Think of Captain Chesley "Sully" Sullenberger, who famously crash-landed an A320 in the Hudson River after bird strikes knocked out both engines. Generations of pilots once knew their planes intimately, translating thousands of fly-time hours to go into problem-solving mode. This will soon be a thing of the past as AI is rapidly adopted by the airline industry. Pilots in trouble will no longer know how to fly on their own, having garnered little experience outside of a passive mode of monitoring computers.

Researchers studying the Miracle on the Hudson River concluded that a major lesson of Sully's heroics is that to be resilient one must be both prepared and prepared to be unprepared.[9] Anticipated situations are controllable; no crisis can be perfectly anticipated. We need both generic anticipation schemes and fast implementation skills to fit the generic schemes into the parameters of the crisis under a time crunch—meaning the algorithmic solution works great, until it doesn't.

A second lesson is that automation lowers uncertainty by reducing variety, diversity, deviation, and instability. The side effect is it also reduces autonomy, creativity, and reactivity. Increasing order, conformity, stability, predictability, discipline, and anticipation may make the systems more efficient, cheaper, and generally safer in the standard, expected environments, but it also makes them less resilient outside of those boundaries when the unexpected occurs. This is the trade-off between efficiency and flexibility. Algorithms promise efficiency. Only humans bring flexibility.

In the 737 MAX tragedy, AI prevented the pilots from overriding the autopilot system. A MAX has a cruising stall speed of 237 km/h. Below that speed it falls out of the sky. When the unexpected happens, pilots have to realize something is wrong, then test what they think is a solution to prove its viability. The initial promise of AI was to free us from mundane labor. But the reality is that AI is rendering sophisticated labor mundane and dulling the intellectual capacity of those who engage in it. Research on the crashes indicates that high levels of automation lead to higher accident rates.[10]

When authorities initially tried to blame the first 737 crash on pilot error, Sully was not shy about pushing back: "Resurrecting this age-old aviation canard...minimizes the fatal design flaws and certification failures that precipitated those tragedies, and still pose a threat to the flying public. I have long stated...that pilots must be capable of absolute mastery of the aircraft and the situation at all times, a concept pilots call airmanship. Inadequate pilot training and insufficient pilot experience are problems worldwide, but they do not excuse the fatally flawed design of the Maneuvering Characteristics Augmentation System that was a death trap."

Right Answer, Wrong Question

The airline industry's warnings about how AI can hamper lifesaving human creativity are dire. Nic moved on to a more optimistic source of inspiration: sailing. "Think of the America's Cup boats. They're crazy fast, and use a host of gliders and aerodynamic soaring, these things that use the wind and multiply its force. With only its foils submerged or breaking the surface at a given moment, most of the surface area of the boats does not sit in the water. They're in the wind, allowing the boats to sail at three times the wind speed and literally ride the wind. I just thought, there must be a way to make wheels that are simple, light, and stable."

This is learning by noticing, by being attentive to the world.[11] It started with the expression of a personal desire, "I want to bike faster." It evolved to a recognition of being an embodied being, and the limitations that accompany her embodied reality, noting something along the lines of "The hardware for biking faster can only be properly used by certain body types because hiding from the wind requires extra weight." But then Nic went to the next step in **BEAM**, thinking about the external resources offered by the natural environment. The wind needed to be re-conceptualized as a potential resource that could be harnessed to support what her body was trying to do, not a liability that had to be avoided. She realized the truth that "there should be bikes, like there are boats and planes, whose increased speed is powered by riding the wind."

Her problem identification was also a business insight. "All of us consumers, all cyclists who are buying wheels, want to go faster. For upgrades, the one question we want answered is: 'Will I go faster?' And up until now, there has not been a single company, or coach, or bike shop, or anyone selling a piece of equipment, that could say, 'Yes, you will.' My whole thing is there had to be a way, something that works for any rider. Where weight, shape, experience level, technical ability, weather conditions, any brutal crosswind, doesn't matter. And there was."

Nic's invention has three distinct innovative design features. The first is a sidewall that allows the wheel to absorb wind gusts. The surface area of the wheel is flexible, forming the most aerodynamic shape for any wind condition, working something like a bird's wing. The second innovation is the way the wheel harnesses the power of the wind by turning crosswinds, which used to be a debilitating source of challenge to riders, into forward thrust. Instead of slowing one down, crosswinds would now speed the bike up. The third distinction is that her wheel is light, disproving concretely once and for all the assumed necessary trade-off between aerodynamics and weight.

"I designed a disc wheel that was lighter than even the lightest wheel I rode before. But what I didn't know was that it could actually produce forward thrust or propulsion. That was an unintended benefit… So that's when I looked into the physics of it all. And once you figure out exactly what's happening and how wind forces can be translated into propulsive forces, which is also a stabilizing force."

She explained that every time she tested the prototype, there was something that didn't quite add up. But she viewed this as a good thing, as she was committed to repeated experimentation and stayed optimistic throughout the emergent difficulties.

> *My way of thinking has always been to solve for x. I envision the end result, and then try things. There was an unseen benefit that I wasn't expecting. And I didn't know what it was, and I also didn't know if I was crazy, because I wasn't really telling anybody about this. I was riding my experiments. And I thought, 'Okay, well what's happening?' Then I would go and do the research and try and figure out what it was. Once I understood the physics of what was happening, I had a better idea of how it should have been designed. So, I kept adding small innovations to my original design, until it became what it is right now. And every single big breakthrough was a happy accident.*

Tinkering and experimenting spark learning through imagination.[12] And given the mix of experience, knowledge, and imagination, it's fair to say that creative tinkering is born of expertise. That means expertise is a necessary condition for creativity, not that all experts are necessarily creative. Even though Nic didn't have the formal training of an expert, she quickly became one through her artful experimentation. Here she was showing the industry that its business model didn't have to be the way it was. That's inspirational.

> *The industry is coming out with products that have the right answer to the wrong question. And nobody from within steps back to recalibrate, "What are we trying to do? What are the goals?" Once this has been established as the direction that equipment is being developed in, you are going to stick with it. You see what your competitors are*

> *doing, and you're going to try and make something better than them, but within the same mold. It's so limiting. It's tunnel vision in the worst possible way... To me, everything was a possibility. I didn't have anyone that ever said to me, "well this is how you have to do it," or "you don't produce it that way." I still had that wide open perspective. And I was totally okay with modifying my own equipment and doing all the things that you're not ever supposed to do.*

I know and respect Nic's parents. There is no doubt that she was blessed from an early age with a nurturing environment that never put constraints on her thinking. It's fair to ask, as we work through a wider version for bringing artful thinking into the mainstream, if we are laying the foundations for empowering the next generation to think boldly like this. Unfortunately, researchers have been warning that our current educational delivery systems can be traced back to "disembodied" views of human thinking,[13] the incorrect belief that our cognitive processes take place exclusively in our brain and nothing is gained by including our bodies, the environment, or physical activity alongside contemplation. Not much has substantively changed since the days of my early education and the message that our thinking had to be freed from the limiting constraints of bodily senses and the trappings of the physical world.

Embodied learning sees actions, emotions, sensations, and environments influencing what is learned. What are things you can do to support better learning in your own life? Small mindful changes like bringing out the notepad and handwriting your thoughts instead of typing on a screen; not being shy about using hand gestures and other types of movement to convey ideas, even mathematical notions; using real-world trinkets you find around you to solve math problems; and not being uptight about love. (I promised that Nic wouldn't bring up the term, not that I wouldn't.) Emotional words have a surprising power to accelerate knowledge acquisition. Algorithmic supremacists' vision of a future where every child is stuck with an AI tutor moves us in the exact opposite direction. We lose mirroring, emotion, gestures, and the benefits of manipulatives in their paradigm.

Nic offers the following:

There are two ways of thinking. One is the constrained, "this is how you do it." The other way expresses that there are different ways to do what you want. The end goal for bikes is as simple as "can any rider go faster?" Can you answer yes to that one question of "will I go faster if I put that on my bike?" It's super general because it's not even limited to a wheel or a frame or whatever. It could be anything. And then beyond that, it should be rideable and faster for any rider in any weather condition on any course. And that is so general that it opens up your mind to being able to trip across anything along the way. Take that fork in the road and see where you go from there. I know what I know, I don't know what I don't know, and that's usually where the answers are.

Nic's philosophy echoes our earlier discussion that the best choice to make is the one that opens up more choices. She lives it in her professional life.

The story of the first time she tested the wheel in a public race is wonderful. It was a 90 km Half Ironman. She hadn't been training for a while and was nursing a leg injury, so she only competed in the biking and swimming components of the challenge. Even so, she rode the fastest run of her life thus far. As she finished the race, and all the other competitors were still at it, one of the timing officials approached her. He was holding in front of him a chocolate milk container because that company was sponsoring the race. Nic explained with a smile, "He brings it over and goes, 'Will you pee in this container for me?' I had raced too fast. They couldn't understand." But, of course, what enhanced her performance was not drugs. It was her innovativeness.

One neat habit she has that helps direct her to that fork in the road is when looking for materials for a project, she deliberately walks down the wrong aisle in the hardware store to kick-start her emergent thinking. "What usually gives you the best start-up ideas is looking in the exact opposite direction. That 180-degree turn takes your mind somewhere new. If you do things the same, you're always going to end up in the same place, with the same products, and no innovation. Innovation lies in something that is absolutely different."

The practical side of artful innovating is defying the algorithm. Whether it's wrong-aisle shopping or wrong-industry modeling, deliberately escape the confines of the rules. And whenever you can, get out of your head and tinker with what is in front of you. Mess with things in the material world. Get physical. Every bit of research and every inventor anecdote tells you that failing to engage with the resources presented in the immediate environment is a missed opportunity. Change is possible and within your power—you literally just have to reach out and creatively engage with what's in front of you. You are not an algorithm that can't proceed when broken. Your mind is flexible and adaptable. Figure out a way forward.

PART IV

ARTFUL LIVING

10

—

The Least We Can Do

MONOPOLY AIN'T WHAT IT USED TO BE

Early in my academic career, I was obligated to lecture MBAs on the economic roots of strategic thinking. This included discussing the infamous supply and demand curves of economic modeling and the equilibrium pricing that emerged from the analysis. For those blessed souls who were never forced to spend time with economic models (you are probably better human beings for the ignorance, less greedy and selfish according to the data,[1] so let me apologize in advance for the coming exposure that may ruin you), the gist of it is that any given market can be mapped with a downward, sloping demand curve capturing how many units of an item customers will buy at certain price points, and an upward-sloping supply curve showing how many of those same units sellers would be prepared to offer at each price. Where the two curves meet is the equilibrium, the pricing sweet spot for perfect buyer/seller alignment, leaving no goods unsold or demand unmet.

I used to explain to my students that models like this are described as "elegant" because they simplify the very complex reality of economic exchange into simple, digestible, easy-to-replicate management tools. After the briefest taste of economic theory, the newly enlightened can arm themselves with charts showing how "X" marks the spot in supply and demand, allowing for an easy-to-follow pricing algorithm.

Yet as we all know, the real world is far more complicated and inelegant; if you were to stand around in a store at your local mall for an hour hoping to buy jeans at a discounted rate as the foot traffic waned, you would be wasting your time. The shopkeeper would not be constantly adjusting the price in real time to find equilibrium. Prices are basically set for an extended period of time, a reality we are grateful for as it brings some sense of stability for all parties involved in the transaction.

Sure, prices may be adjusted on occasion for seasonal promotions and special sales, and in some spaces, there is even room afforded for haggling, but in general both seller and buyer can plan with confidence based on posted numbers. It's a system that is economically inefficient, and somewhat messy, but it's allowed markets to function just fine for centuries. I let students know that they are not going to find normal people chasing precise demand/supply efficiency in the real world—that would be an ugly, chaotic, and antihuman sort of market.

Well, it's a good thing I've retired those slides because, since then, algorithmic supremacists have taken a shine to dynamic pricing. It started with Uber and surge pricing for rides during periods of high demand and difficult driving conditions, an easy tweak to the company's algorithm given that the buyer/seller exchange took place exclusively on its proprietary digital platform and not in a face-to-face confrontation. By the time the driver showed up, riders would have already made peace with the manipulative pricing.

This practice then crept over to Ticketmaster. That company had to make a few changes to its operating model before this inhuman practice could be implemented, like eliminating first-day in-person sales at box offices. To facilitate real-time algorithmic adjustments to pricing in the frenzy of a high-demand onsale, all parts of the transaction had to happen in digital space. It wouldn't work in the more social, even civil, atmospheres of people lining up at local theaters or getting their tickets at record stores where human decency and respect for culture still mattered.

Now it seems that executives at Wendy's are hoping to bring the madness of pricing efficiency to their restaurants.[2] During a February 2024 earnings call, CEO Kirk Tanner announced the company's intentions to invest $20 million in digital menu boards that will be algorithmically enhanced to change prices in real time based on how busy a specific location is at any given moment. I can't imagine how dreadful this rollout will be, should the plan actually be executed. As if it weren't bad enough that hardworking folks are stuck lining up for an affordable meal during a very limited lunch break, they will now get to watch the price for their meal rise as their patience wanes and time disappears. A Wendy's spokesperson added that "dynamic pricing will also enable discounts at slow times of the day," but the "also" qualifier lets you know what the primary intention is.

Dynamic pricing works best in monopolistic, or near-monopolistic, conditions. All rideshares have adopted surge pricing, meaning if they have successfully disrupted the preexisting taxi industry in your market, you have no choice but to adjust to the new normal. And Ticketmaster only gets away with these exploitative practices because of its near total lock on the industry. My guess is that if Wendy's actually goes through with this, it will be most successful in geographic locations with few peak-mealtime alternatives.

That was another part of my lecture—that economics-infused approaches to strategy love monopolies, like the famed Five Forces framework created by Harvard Business School's Michael Porter.[3] Apologies again for making you a worse person if you've lived free of the 5Fs, but the algorithm is that within any industry a company's competitive position is threatened by the relative power of buyers and suppliers, competitors, substitute products, and threat of new entrants.

The game of strategy then becomes enacting tactics that will minimize the power of these forces. If you play the game well, achieving maximum efficiency and effectiveness in crushing the will of the human beings that trade with you (sorry, I mean forces), the prize will be a monopolistic situation where customers and suppliers have no power to push back or bargain, direct competitors and those with substitutes outside of your industry are irrelevant, and no new competitor can overcome existing barriers to entry to join the game as you own the board, the pawns, and everything else.

Michael Porter, I imagine, would only take issue with one fine detail of this telling in insisting that governments in capitalist democracies would prevent monopolistic outcomes by legislating strong antitrust regulations and aggressively going after the players who don't respect competition. How's that been working out? Not only have Big Tech's monopoly-like powers grown, but they're also proving to be very bad at the activities they were good at prior to becoming monopolists.

Google's monopoly, for example, isn't working out that great for us... or them. Writer Corey Doctorow has been documenting the "enshittification" of the internet. He observes that society made a deal, of sorts, with Google, which goes something like this: "You monopolize search and use your monopoly rents to ensure that we never, ever try another search engine. In return, you...promised us that if you got to be the unelected, permanent overlord of all information access, you would 'organize the world's information and make it universally accessible and useful.'"[4] But Google has broken that deal. Its search results have become terrible, a haven of AI-generated garbage, SEO-optimized spam, and clickbait scams.

How do we live artfully in a corporate-dominated economic

environment where everyone wants absolute control, businesses are run on algorithms despite the human costs, and nobody seems to be making things better even with access to unprecedented wealth and power? This is the question I posed to Douglas Rushkoff, an intellectual who has written deeply informed and novel takes on topics as diverse as technology (he coined the term for a media creation going "viral"), business, economics, and even religion, spanning the realms of nonfiction, fiction, and even graphic novels.

Rushkoff sees in the worldview of Big Tech a defeatist take on human potential, innovating destructively in the near term while hoping for longer-term innovations to emerge that will allow them, but not the rest of the world, to escape the social, environmental, and economic consequences of the disasters they have created. He labels this "The Mindset,"[5] explaining, "They are preparing for a digital future that has less to do with making the world a better place than transcending the human condition altogether. These billionaires once showered the world with optimistic business plans for how tech might benefit human society. Now they've reduced technological progress to a video game that one of them wins by finding the escape hatch."

How do we continue to live as creative and productive beings alongside these destructive forces? By rejecting The Mindset. Douglas laments, "We've all participated in The Mindset, even if it was only to believe in the inevitability of our own victimhood." He's right. Be honest. How many of us have chosen to either mirror The Mindset through a defeatist attitude, or rebel in a way that affirms it by putting our humanity second and playing their game?

Google is failing us in search. It was also deeply embarrassed by the disastrous 2024 launch of Gemini. Much of the tech world was hyped for Google to finally release its AI; designed to meet the promises of MUM; learning from the mistakes of its immediate predecessor, Bard; and using the unmatched resources of this global behemoth for training data and sheer computational power. Except, the lesson learned from algorithms that discriminated against women and minorities is to discriminate against men and Whites. Gemini was unable to generate images of White people, even when asked for historical figures.[6] It put Black folks in Nazi uniforms.

Gary Marcus feels Gemini never stood a chance as Google's team were rushing to get an AI to market that could balance cultural sensitivity with historical accuracy. But the tech just wasn't there yet, and who knows when (and if) it will be. It would need a system "that can distinguish past from future, acknowledging history while also playing a role in moving toward a more positive future...[and] a nuanced understanding of what is even been asked."[7] We have none of that, and yet Google is planning on fast-tracking Gemini into all its software, platforms, and devices.

Google's executives want to exploit the high switching costs of abandoning their ever-worsening products, leverage their success in either kneecapping or acquiring upstart competitors and their airtight relationships with others in the Big Tech cartel, and live their best monopolistic lives. They aren't counting on you using Google search because it's the best; they are counting on the fact that either you've bought their expensive hardware and are stuck, or that it remains the default on Apple devices and you are lazy.

But there are ways to create some distance between yourself and Google. Gemini need not inevitably be part of your intellectual future. You can start making informed choices. You can stick with Gmail but ignore the AI's autocomplete functions so that you stay in control of your own words. Make your choices on a product-by-product basis knowing that an absolute requirement of innovating ethically is that the innovator be motivated by a sense of hopeful optimism in the future creative contributions of fellow humans. You can use the Google Maps app on your phone, because that product exudes the optimism of exploration, but not download Gemini, which has clearly been designed for social engineering. Hold on to the hope that better products will arrive, and don't be a mindless consumer of tools that undermine your needs.

What Are You Gonna Do with My Money?

There are alternatives to the algorithm of monopolistic control that are still pro-innovation. It seems that in the MBA world, where I've spent much of my professional life, a set of blinders comes with the uniform. The last major creative reconceptualization of what a capitalist economy might look like was offered by my mentor R. Edward Freeman with his stakeholder book,[8] released in 1984. Even as so much of the reality on the ground changes, the discourse of what an ideal market economy looks like remains static.

Rushkoff has found inspiration in the actions of those who have creative strategies for ensuring that money circulates within their community. "When you create a boundary condition around your economic or business model, you can get a sort of Dyson vacuum cleaner effect, and you get circulation of money. What is it like to earn $1 ten times, rather than earning $10 once? I love the example of US Steel Workers, who took their retirement funds and invested it in projects that hired steel workers, and then invested in retirement homes for their own parents, and again hired steel workers to build it. They get this triple whammy with their own money."

This should be on the slide of every Strategy 101 lecture: instead of viewing economic partners as abstract forces who hold power that you need to crush, view them as fellow community members you want to support in win-win outcomes. Sure, there is a pinch of tribalism and ideological homogeneity in this recipe, necessary to keep the mixture smooth and trusting, but there's no reason why it can't be universalized for all tribes and all communities. Again, Big Tech looks at our idiosyncrasies as variables that can be exploited and homogenized. Google wants

Gemini AI to make sure we all think the same. Ticketmaster wants to ensure that only the highest payers get to fill arena seats. Wendy's wants to thin its lineup by charging a premium for its generic burgers. Why can't we look at the world and say, "Forget dynamic pricing—I want the steelworker's triple whammy for my community!"

It comes back to approaching business through the **BEAM** lens of embodied beings acting in physical environments with limitations, not as abstract or elegant systems of efficiency maximization. As Douglas explains, "When the abstracting forces of capitalism, digitalism, and individualism all dovetail, we end up trying to turn physical reality into an infinitely scalable symbol system. We do it because we're so afraid to recognize our limits and work within them, which actually promotes circular prosperity and renewal."

Some business elites respond with hostility to his pitch to round the economy. They don't get it. Douglas laments, "But when I talk about that at conferences, businesspeople say 'Wasn't that illegal?!?'" It's simply wild that executives who see no conceptual problem with changing the price of a burger, concert ticket, or ride every few seconds as you wait in line for your turn to be served judge efforts to keep money circulating as morally problematic...another symptom of the poisoning effect of an economics education. To them, dynamic pricing is the digital realization of equilibrium pricing finally come to life. It's progress. It's elegant. This is something different.

Douglas sums up the circulate alternative as "Doing things where you're going to get to see the money again, rather than converting the money from kinetic energy into potential energy. Why do you want to store it? You look at our economy, and it's all batteries and no juice. That's a shame." The message here, which is sound advice for all of us, is to make your investment decisions knowing that you'll see that money again, that your partners will support you as you supported them, not hoard the wealth.

But as we noted earlier, Big Tech has also gotten wise to a version of this. Microsoft's $10 billion investment in OpenAI must be spent on Microsoft services, and Google's $2 billion to Anthropic must be spent on Google Cloud. They are learning how to circulate their money. Big picture, though, we can find a silver lining in this mind shift. Normalizing circular economies can be a good thing, so long as **intent** takes on an outsized importance. Always scrutinize motives before you hand over a dollar, demand accountability and clarity to justify your trust, and only work with those defending an authentically hopeful take on humanity.

For years I sounded the alarm on executives taking an agnostic stance on intent. Too many make the cynical claim that intentions and explanation are unimportant, and we should happily take the win anytime corporations are talking about something other than profits. But the differences between intentions are everything. Moral dilemmas get resolved by how we defend our **VICE**.

Yes, that means that it is the arbitrary, inconsistent, irrational, and unpredictable external assessments made by the folks we are trying to convince that matter most. There's no consolation prize for being hyper-rational or meticulously following the algorithm of reason if you haven't brought the community onside. This doesn't mean that artful intelligence gives up on truth; it just means we are once again mindful of limitations. Having truth onside has not always served the holders very well.

Artful inventions should be crafted by innovators whose justifications celebrate human potential. If we don't want to limit the scope of innovation itself, we need to focus on the moral intentions and explanations of the innovators—how they defend their **VICE**, not just ship products. Why does corporate intent matter so much?

Well, in the 1990s, BP's push for more restrictive and costly environmental regulations, a departure from industry norms of the time, was lauded. However, the intent behind its pro-regulatory stance was to increase costs for rivals. And the Deepwater Horizon oil spill in 2010 showed its principled commitments held only when financially advantageous. In the following years, it was revealed that BP was responsible for the oil spill because of deliberate misconduct and gross negligence. Worse, the company offered only a cosmetic solution to the cleanup, using Corexit to visually cover up the damage, not doing the work an authentic commitment to environmental responsibility, or even honoring the regulatory requirements, would demand.[9]

Actions divorced from intentions tell a misleading story. BP's brief love affair with strict environmental regulation was a story of causing competitors pain. It took time, but the nature of its executives' nasty intentions came out. And sometimes the reverse occurs; intentions that seem nasty on the surface may turn out to be of a different sort. For example, in 2023, antisemitic social media influencers took a pronounced interest in a news item claiming that a rabbi had bought Pornhub.

To clarify, the individual in question was ordained a rabbi, but that's not his primary identity, and it was his private equity firm that took a controlling interest in MindGeek, the parent company of these sites. But in his own words: "When a potential MindGeek acquisition became available, we took an open-minded, unbiased look at it. What we found absolutely blew our minds. We found a company that had created the best-in-class online trust and safety automation tools… We're of the belief that these tools must be shared with the larger internet."[10] Intentions, and explanations, matter. We can't judge moral character by only knowing what people do.

End of Certainty

Douglas shares the story of an uncomfortable exchange he had with evolutionary biologist and algorithmic supremacist Richard Dawkins at an exclusive NYC party.[11] Dawkins proceeded to mock him for believing in a "potentially moral universe," as the assembled dignitaries chuckled. Rushkoff was arguing a position I am very sympathetic to: that the universe leans toward morality, trying to make the point "that morality is a thing, that there's more going on here than just competition between genes for dominance." The hyper-rationalist Dawkins considered Rushkoff to be a victim of delusion for holding a belief in the necessity of ethics.

This certainty is a trait we've been highlighting in algorithmic supremacists throughout the book. It's not just that they are determinists seeing humans as automatons; it's that they hold this belief with the unflinching conviction that anyone who explains life differently is irredeemably delusional and not worth their time. To algorithmic supremacists like Dawkins, only he and other scientists were gifted with the ability to see the world as it really is.

> *He couldn't acknowledge that his own commitment to scientism is based on something...more like faith in an empiricist universe... his insistence on living in an evidence-only universe isn't based on evidence at all. It's an assumption. It's part of a system of meaning, developed by a community of people over time. It just happens to be a meaning system that ignores meaning itself. Worse, by rejecting the validity of any other meaning system, it is prone to instilling in its adherents a sense of superiority over others.*

Unfortunately, we find ourselves in a political moment where rejecting the validity of other meaning systems is a required prerequisite to obtaining a public platform. Consider the algorithm currently in vogue for political allyship. To be an ally in the fight against oppression, a complicated series of if/then steps must be followed. The algorithm starts by demanding a commitment to recognize all sources of personal privilege, including that which may emerge from gender, class, race, or sexual identity.[12] But privilege is also defined as "characteristically invisible to people who have it...granted to people in the dominant groups whether they want those privileges or not, and regardless of their stated intent."[13]

Intention, in this case, doesn't matter. Our explanations don't matter. Our ability to know ourselves is an impossibility, unless guided by someone who can see what is invisible to us. These premises undermine **VICE** completely. As I've written elsewhere,[14] by discouraging all but the most untainted by privilege from trying to help in the fight against oppression, the new allyship algorithm makes it harder to stand up for others and confuses the historic language around what it means to be privileged or oppressed.

Demanding that would-be helpers pass a purity test is not the only way to build a better, more empathetic world. Advocates of allyship assume that the natural inclination of would-be helpers is to perpetuate systems of oppression. The allyship algorithm is just as likely to give rise to an uncooperative society as allies are expected to speak less and hold back their ideas and opinions or, worse, become resentful and rethink support. This algorithm, like all algorithms, is many things, but bias free is not one of them.

I was surprised (but not), to learn that Rushkoff had whole chapters drafted for his *Survival of the Richest* book observing something similar that got cut before publication. His original intent was to include a section critical of the way "the social justice warrior class and intersectionalist academics" were as committed to a dangerous algorithm as the billionaire class. Both are interested in control, thinking that their in-groups can somehow operate independently from society at large, with no need to compromise or find a way to create space for those they disagreed with. Douglas noted, "It echoed Richard Dawkins, and the idea that what we intend and how we feel, and our perception of reality doesn't matter."

The ultimate emptiness in the discourse of intersectionalist academics, and their complete detachment from the thinking of regular people, was brought into troubling light recently when the presidents of the University of Pennsylvania, Harvard University and MIT were summoned to testify to the House Education Committee and could not bring themselves to assert that calling for the genocide of Jews on their campus would be a violation of their code of conduct.[15] Instead, they responded, some with a smirk, that it would depend on the context. Institutions with massive administrative infrastructure solely in place to encourage the purity of untainted allyship, dismantle oppression, and snuff out privilege needed a contextual analysis to determine whether a genocidal call, in this specific case, might be a bad thing.

These ostensibly liberal individuals could not bring themselves to think, and answer, like humans. They were locked in the algorithm of allyship, and the algorithmic path was not leading them to a liberal certainty. Some believe the step where the algorithm broke down was in the DEI calculus of powerful versus oppressed, and the context they needed was to determine whether a call for genocide of Jews would be an acceptable targeting of the powerful. In this system, Jews are defined a priori as a group that cannot be a victim.

The deaf-to-history algorithmic code doesn't allow for the human understanding that the context requires. Others mused that the calculus went awry when the context was foreign students making the genocidal statements. Should the school's leadership decide this class of students was in breach, these individuals would lose their funding and likely be deported, and thus were the more vulnerable population in need of protection, since it was deportation in action versus genocide in words.

There have been some apologies for the sentiment but no clear statement of the thinking that led to such ugly displays. Ironically, in this instance, the chatbots employed a morally superior algorithm to the human university leadership. In the hours after the testimony, social media influencers documented them asking the same questions of ChatGPT.[16] It answered, "Promoting or calling for genocide is not only considered harassment and bullying but also a serious violation of ethical and legal standards. It goes against principles of human rights and can lead to severe consequences. Such actions are not acceptable and are condemned globally." It really should have been that simple.

Closer to home, the algorithm of oppressed/oppressor, the redefining of words, and an oblivious disconnect from the harms it was causing on the ground took a disheartening turn at my institution. A group of colleagues, including a tenured professor at the university I work at, vandalized the flagship Indigo bookstore in downtown Toronto, falsely accusing its Jewish founder and CEO, Heather Reisman, of "Funding Genocide."[17] To add to the pain, these so-called progressives decided to take the action on the anniversary of Kristallnacht, a night of broken glass for thousands of Jewish businesses in a Nazi-led pogrom just before the Holocaust.

These committed allies were arrested by the local police[18] and suspended with pay from the classroom by the university administration. You may be wondering, how did a majority of my liberal university colleagues respond to this development? With a walkout of class, supporting the vandals, in the name of allyship and justice. They sent out a statement, demanding "the reinstatement of our colleagues...unilaterally placed on leave because they have been accused of engaging in peaceful protest, an accusation that carries with it the even more spurious charge of being 'hate-motivated.'"

In their algorithm of allyship, all action taken in the name of the oppressed against a perceived oppressor must be viewed as necessarily peaceful and just, not violent and immoral. "Words are violence" became "violence is speech." More shockingly, the statement added, "We expect, at a minimum, that York University will respect due process, including the core value of the presumption of innocence" and "We categorically reject the logic of community safety that serves as a rationale for the suspensions of our colleagues." Safe spaces are demanded for those calculated to be at the top of the oppression hierarchy, but "the logic of community safety" is abhorrent when employed by a perceived oppressor.

The statement ends with a repetition of all the inputs to the algorithm that the mission of the university is committed to: "cultivating the critical intellect," "exploring global concerns," "academic freedom," "social justice," "learning environment committed to the public good," "progressive approach," "diversity and inclusivity," and "passionate about advancing social justice and equity." All these inputs, all these words—they used to mean something everybody could understand. Now, they are being employed to support an end that sensible people find criminally problematic.

I spoke out when the walkout was announced, curious as to why the solidarity for those who "dare speak up" was narrowly limited to those in lockstep ideological alignment but broad enough to include "speech acts" so uncivilized that they met the high threshold for arrest with a hate-motivated qualifier. I was also thrown by "the presumption of innocence" line. Forget the fact that these individuals celebrated and publicized their "activism." Does my union no longer respect the rule of law and the processes set in place to protect students after a charged breach in the rule of law took place? If a professor was caught abusing a student, should that person remain in the classroom until the judicial process has run its course?

The response from a representative was "We have a colleague who has been arrested for an action done outside the university… Solidarity with those who dare speak up against injustice and oppression, whatever form it has, wherever it happens. We shouldn't have a selective mode of what we stand for and what we turn a blind eye on." I asked how about, as an alternative, we offer solidarity with those seeking to create a university space committed to harm reduction and minimizing trauma?

For a student already suffering in this hostile environment, seeing faculty go to bat for someone charged with a hate crime is traumatic. The decision to discount the harm caused by normalizing hate crimes is, in fact, an injustice and an abdication of what it means to be professionals who work with ideas. The representative responded, "Hate crimes shouldn't be normalized, but I don't find any rationale of equating being critical of a country/government/army and their supporters with harboring hatred based on religion, ethnicity. The latter is against the law and should be punishable, the former isn't and should never be in a democracy."

I wish I had seen the cut chapter that Douglas had crafted. The malaise of algorithmic supremacy is nonpartisan in its affliction. It's hitting Right and Left, capitalists and socialists. But it's that same idea of putting all of humanity into limiting and uninteresting boxes. Douglas clarifies that "it's the collapse of nuance in a digital media environment that *is* the problem. It's the binary, yes or no thing. And you're right. That was why I wanted to have the Social Justice warriors in my last book, because I wanted to show that even when one side is more right than the other, both sides are inflicting us with a stridency that hurts everything."

But he reminds me of a reason to stay hopeful: "That's why I keep turning to Team Human... Life itself exists in that squishy place between the other things, between the quantized notches of the digital world. The digital is the ticks of the clock. Which tick are you on? I'm not on the ticks. I'm in the duration of the seconds between the ticks, when time actually happens."

To operate in between the ticks is to retain your sense of language, your sense of decency. I don't believe that the presidents of those universities or my colleagues here were being insincere. But I also don't think they were doing the work of defending their **VICE** either. They are so committed to the algorithm that they think it's enough to just throw out terms like "context dependence," "justice," and "free speech" to defend doing things most of our society views as immoral. If you're going to defend the immoral, you need to do so explicitly, in your own words. But don't think that we will simply march in lockstep with you just because you have certainty. Whether you are a billionaire tech exec, a millionaire Ivy League university president, an upper-class professor, or a working-class activist, stridency and self-satisfaction are no substitute for the messiness of being called to account.

Awe Inspired

Staying human in the overbearing age of surveillance capitalism demands incremental mind shifts. Artful work is often done in small steps, winning people over one conversation at a time, not big, outlandish, difficult-to-implement solutions. Douglas believes we need to get people thinking about the difference between the preexisting conditions of the natural world and the synthetic rules and value statements, like the corporate power grabs, that we've been led to believe are simply part of the laws of nature. Nothing is wholly inevitable or static. Much as with the bodily limitations discussed earlier, people can accept boundaries on what might be changed while still being energized by the possibilities that are within our reach.

"So, what can we do about it? That's the tricky part. When the Internet came, I thought, everyone's gonna make their own videos. And of course, on YouTube, or any of those platforms, it's like 2% of people actually make videos and 98% of people watch them. Even when a new thing comes along, I was liking TikTok at the beginning because my daughter was actually making those dance things. But now she just swipes."

The initial promise of the internet was social mobility—suddenly, the power to access information and to message millions of people was democratized and made accessible. The barriers to building a business, showcasing your skills, or finding partners to help bring your innovation to market were dropped. In theory, all you needed was the drive to be creative. But Big Business quickly populated the Web, controlling who gets seen, and the platform became yet another outlet for monopolists. Then social media emerged, and once again, it briefly seemed like the power was back in the hands of creators. Until algorithmic supremacists used their innovations to pull the levers of control and encourage passivity in their user base. Where are the ethical innovations to reverse this trend? Who is going to step up and design future products that will support creativity and social mobility? This is what we should be demanding. If you must be on social media, choose to post, not to swipe. What else?

Douglas imagines a future where it is easier for people to ask each other for favors, a first step toward better human connection and a shift in the innovator mindset. He wonders why so many of us are reluctant to borrow a tool from a neighbor instead of buying a new one. Another simple prescription for thinking more like a human: try sharing and cooperating before you choose purchasing or hoarding. If you ultimately need to go it alone, that's fine. But do your part to strengthen communities of mutual support.

While digital technology often undermines these social mechanisms, real world face-to-face contact recalibrates our nervous systems and establishes rapport. Let's work for a shift from a consumerist, individualist, or transhumanist bias toward a humanist bias. We are far better off trying to mitigate the unknowable effects of what we do by choosing to trust and cooperate with other humans who may be able to do things we can't. This is a risk worth taking. As you create with current partners, always be open to finding new partners down the road.

Both Douglas and I have been talking a lot about awe lately as a key step on the path to strengthening Team Human. Psychologists Dacher Keltner and Jonathan Haidt identify awe as a little studied emotion situated in the upper reaches of pleasure and on the boundary of fear, a complex, positive, and self-relevant emotion characterized by perceived vastness and need for cognitive accommodation.[19] Awe impacts the perception of self, reducing its perceived significance.[20] Interestingly, while most positive emotions have an arousing effect on embodied humans, awe has the opposite effect, reducing sympathetic influence on the heart and keeping us still.[21]

I shared with Douglas that when I've given talks to fellow scholars, they get excited by the possibility of awe as a tool. And when I speak to tech executives about awe, they too get excited by the idea...but the outcome of their awe-inducing journeys doesn't lead them to greater humanism. Rushkoff saw something similar: "They go to South America for an ayahuasca trip, have the experience of awe, and come back with the realization that they're God!" He continued, "You know, I think people feel safer doing awe in a religious context. If we say, 'without an experience of God, we're fucked...whatever you think God is...'that's the play."

It's an interesting point. My last book explored spirituality as a point of entry to bring meaning, connection, and awe to the business arena. And much of the research undertaken for this project found surprisingly religious motivations in the mindsets of algorithmic supremacists. What's the piece of a religious experience that is so important for thinking like a human? Douglas mused,

> *People have church, really, because you need to be able to say the worst thing you've done and have it be okay so you can move on... if you're not Catholic and you don't want to confess to some priests, you have to find someone else to do it with. And the Twelve-step programs are, in some ways, for that. Here's a community of people who are as fucked up as I am, in the way that I am. With them, I can unburden and absolve. A bit of anarchism, you know, because you're having the people do it, rather than the priest. You did it with your community. People need that. And that's not necessarily an experience of awe, but it's similar in that it requires the acceptance of a higher power and recognition of one's own unworthiness. We are all sinful, we are all fucked up, and you've got to accept that and speak it out loud in order to move forward in life and be useful.*

Feeling a sense of awe is the antithesis to the economy of certainty that surveillance capitalists are trying to build. It's recognizing your smallness, even for a moment, and having your frame of reference rearranged to accommodate the fact that you are surrounded by things bigger than you. Douglas continued, "What those tech executives really need is someone saying to them 'You are not God. You are just a flawed individual like any other person.' We all need this experience, which is, by the way, anti-intellectual." It's true. This humbling is something that needs to be felt in the body. It is not a rational argument that can simply be processed in the mind. You can always calculate a way to overvalue yourself until you mindfully *feel* like you are part of something more.

Douglas confessed to coming to this point of view recently, a result of an intellectual shift away from an earlier contrary position. "It's the opposite of what I used in my book on Judaism, *Nothing Sacred.* I used the Talmudic argument that if you're in a new town, find where the synagogue is, and walk to it. But if you find out where the *bait midrash* (house of study) is, *run* to it." This advice is very much rooted in the dualism I shared at the start of the book when reflecting on my Yeshiva years. The highest good was refining the soul through the intellectual act of rational study. Sure, prayer is a necessary part of religious practice, but it's a service of the heart, not the head, and therefore seen as secondary. Douglas explained, "And now I realize, no, the only way most of us can experience awe is not intellectual. It's in the safety and the comfort of this other thing. For most people it's not going to be going to the Amazon and doing ayahuasca. It's going to be going to Taylor Swift, or Phish, or church, or whatever. It could be a family singing around the piano. People used to do that."

It's so interesting to see how we keep coming back to love and music. Music is where most of us are likely to find awe, whether in the concert hall as audience members, in the pews of churches as parishioners, in the pub as revelers, or around the piano as party guests. When Nels Cline and I schmoozed on this topic some years back, he called it a "wordless consensus." He similarly lamented the slow vanishing of a willingness to regularly be vulnerable in a crowded room of strangers, a common state of being in our religious past. We experienced a wordless consensus when we would gather in a space of worship to sing together. We allowed ourselves to be fully present and participate in the creation of something that was so different from everyday interactions that we labeled this type of singing "sacred."

Experiencing awe together, brain waves synced, as opposed to having a solitary experience, is on a whole other level. It's drawing strength, as Douglas describes it, from the recognition of common vulnerabilities. This conscious feeling is something that a lot of us no longer fully experience in our tech-infused lives. So, make an effort to gather around a campfire with a guitar, be part of a choir, or take in a concert. See what it does for you when you have the intent to experience awe.

Move Up

The best way to build as we climb the ladder (whatever ladder it may be) is to raise people up with us. Unfortunately, Industry 4.0 has thus far shown itself to be characterized by a generation of downward mobility. And those at the top are disinterested in keeping everyone moving. They are just afraid of the violence that will come when the bottom rung gives out. Algorithmic supremacists pat themselves on the back for crafting policy papers arguing that politicians need to institute a universal basic income (UBI) as their tech takes away the jobs of creative humans. They want governments to provide an income for people precisely because these tech leaders envision a world where their machines will be doing the work people used to do. All profitable, value-creating labor will be theirs, and theirs alone, to exploit. But they are kind enough to ask that governments ensure the regular folks they displace do not starve.

To defend volition is to insist that all people be free to earn and create economic value. We need more earners, not fewer. The philosopher

Richard Rorty got it right when he explained "the social glue holding society together is little more than the consensus that the point of social organization is to let everybody have a chance at self-creation to the best of their abilities."[22] Douglas adds, "Commerce is good, you know. You just don't want the production of money or the hoarding of capital to get stuck anywhere...You don't want giant, large, foreign corporations running things." What we want from our economic system is a "social reality that allowed for the creation and exchange of value between real people. Not everything goes on the ledger. Many of our exchanges are purely social, even if real value like food or services are offered."

Earlier, we presented the five-step hustle, which empowers those who caused and profited from the problem in the first place to fix and profit from it once again. Rushkoff has also criticized the modern idea that all solutions must make money. It's an idea that's so deeply ingrained, even in the business ethics mindset, that when I challenge executives to rethink it, I get "but I'm on the good side of things...my intentions are ethical." Douglas agrees: "They don't even see how deeply problematic it is to think of every technical solution as necessarily profitable. And not just to make money...I can go make money and make the solution work. What they want is not just to derive revenue, but make money as a scalable business, as growth. You're not even allowed to make this much money and stop there. You've got to be like Pepsi and keep growing. And that's when it's like, oh, you really are stuck."

To close our discussion, I asked Douglas to share what gives him hope. His answer speaks to the power of artful thinking:

> *I find hope in accepting that even if everything is ending, even if it's in ten years from now, there is some part of the human organism that is still here right now, wanting to explore and embrace what it means to be a living, conscious, feeling thing. For me, it's okay that even if the tree is dying, which it may be, there are these leaves that*

are here. And I'm gonna be here to nurture that. And who knows, the better we nurture that, the more possibility there is for the tree to make it through another season.

Being here to nurture life…it's the least we can do.

11

Throwing the Right Party

TAKING CARE OF THE SOCIAL ENVIRONMENT

Writing a feature for the *Atlantic*, Derek Thompson immersed himself in data, getting his hands dirty while working through all the studies that tried to get a handle on America's social fitness.[1] He concludes that we are not a healthy nation right now. The problem is that "Americans have collectively submitted to a national experiment to deprive ourselves of camaraderie in the world of flesh and steel, choosing instead to grow (and grow and grow) the time we spend by ourselves, gazing into screens, wherein actors and influencers often engage in the very acts of physical proximity that we deny ourselves. It's been a weird experiment. And the results haven't been pretty."

Psychology professor Jean Twenge tells Thompson in the article, "There's very clearly been a striking decline in in-person socializing among teens and young adults, whether it's going to parties, driving around in cars, going to the mall, or just about anything that has to do with getting together in person."

None of this, of course, is surprising to us. We've already noted with disappointment how so many are spending less time with fellow embodied humans and more time with screens, algorithms, and AI companions. We're quickly reaching a crisis point as these technological turns exacerbate a phenomenon identified by political scientist Robert Putnam a quarter of a century ago.[2] In his book *Bowling Alone*, Putnam concludes that as Americans choose to spend less time taking part in person-to-person social interactions, civic engagement and the overall health of democracy decline.

Thompson tells us that an unnamed acquaintance defined community as "where people keep showing up." He wonders where people keep showing up today. It's not at churches, temples, synagogues, community centers, parks, sports fields, or even offices as working remotely becomes a new norm. "America is suffering a kind of ritual recession, with fewer community-based routines and more entertainment for, and empowerment of, individuals and the aloneness that they choose... Face-to-face rituals and customs are pulling on our time less, and face-to-screen technologies are pulling on our attention more. The inevitable result is a hang-out depression."

Our productivity, creativity, and ability to build depend on having spaces to gather in. Who is going to build spaces that are better suited to support our unique cognitive and bodily dispositions? Who will figure out how to draw us away from screens and back into social spaces? Well, dear reader, you are. Before wrapping up, we have one more topic to explore: how to willfully and intentionally create environments—physical, real-world, social spaces—that are not only amenable to, but nurturing of, artful living.

Creativity is partially defined as an act arising out of a perception of the environment. Figuring out how to build artful social spaces is not that interesting to those who idealize the trope of the solitary genius. Unfortunately, this group has dominated creativity research for far too long. Perched atop their soapboxes, these voices urge us to create alone, reserving praise only for the lone wolves who do.[3] They justify their prescriptions with findings that suggest individual ideation is superior to ideation in brainstorming groups[4] and that individualistic cultures promote creative expression more than collectivistic ones.[5]

But they are simply spinning numbers to tell a confused tale, another misguided branch of algorithmic supremacy. It's hyper-rationalist researchers who think they've landed on the singular formula for creativity, absolutely certain of the universal applicability of their models, while remaining blindly detached from the realities of living as creative beings.

Breaking with the mainstream orthodoxy within his field, psychologist Vlad Glăveanu argues that, in fact, creative minds are social and cultural, meaning these celebrated acts of seemingly solitary creativity are only made possible by sociocultural relations and interactions.[6] Despite their claims of rigor, the studies we quoted earlier, which claim to prove that thinking on your own is better than brainstorming in groups, or that individualistic cultures are more creative than collectivistic ones, overlook the social contributions to creativity because of a choice: focusing their research on the *end* of a process. They are describing the final moment of discovery, the resolution of working through a problem that started long before an individual creator may have closed the door to their isolated lab or studio, later reemerging to announce a breakthrough. The extensive creative process that brings one to an aha moment is always deeply social, even if the credited innovator won't name it this way.

Glăveanu offers an observation on the ethical implications of his take that reaffirms the importance of the principles outlined in the last chapter: "Sociocultural theories of creativity...emphasize the ethical responsibilities... If difference is at the heart of creativity and taking the perspective of others is a fundamental part of its processes, then...the tendency in today's fragmented and polarized societies to judge rather than understand...endanger[s] our very existence as creative beings."

A Place Where Accidents Can Happen

Glăveanu carves out a space for accidental or serendipitous innovation. "Serendipity" describes discoveries that occur "at the intersection of chance and wisdom."[7] To an individual innovator, the serendipitous moment is marked by a surprising insight. But there is much more going on behind the scenes. Earlier in the book, we shared how Velcro was discovered in a moment of serendipity. There are other examples. Sticky notes emerged from a failed effort to make a super glue.[8] Instead, Spencer Silver, a scientist working for 3M, found he had created an adhesive that stuck lightly and didn't bond tightly. It took six years for him to figure out a use for this substance.

In the cultural arena, the *Grand Theft Auto* (GTA) video game franchise also emerged serendipitously. The original intent was to create a classic cops and robbers–type driving game. But a glitch in the programming algorithm caused the cops to aggressively chase the players off the road.[9] Early testers of the game loved this glitch, and a new type of socially dysfunctional game play was born. Over 400 million units of the game had shipped as of the summer of 2023.[10]

The category of serendipity includes "the unplanned building of social networks."[11] Critical to the creative process are the people we meet unexpectedly, those who turn out to be sources of valuable knowledge or who lead to further valuable connections. For GTA's programmers, it was the testers. For Spencer Silver, it was Art Fry, a colleague who was getting frustrated by the fact that the little scraps of paper he used to mark the hymns while practicing with his church choir would fall out before the Sunday service. He needed a new type of bookmark, one that would both stick to the pages of his hymnal and not damage this sacred text.

Gary Fine and Jim Deegan observe that "serendipity goes beyond the cognitive: connections made between people, or one's being present at the right time and in the right place, can just as well lead to valuable discoveries." We build artful spaces that support those who gather to be open to the feelings of curiosity and surprise, and we go there to take advantage of the serendipitous opportunities. We'll soon be talking about how to do this, but in terms of what's already been shared, an easy start is working in a space that has open windows so you can see the sky. So is moving your workspace from a solitary environment to a more social setting.

Researchers explain that "intentional...serendipitous actions," like mindfully being open to surprise, lead to "serendipitous relations."[12] And because most of us are not wired to fully realize the value of these type of events as they unfold in real time, but piece the connections together later after reflecting and considering the positive outcomes, the ability to be open to the unexpected is a skill that can be developed. One way to refine this skill is through **VICE**, putting yourself in social situations where all gathered are engaging in a flurry of novel, even outrageous, explanations and justifications. I can tell you that spending time with Perry had this effect on me.

In our conversation, Rushkoff mused, "I think it's the social and collective. It's not getting an individual or helping encourage an individual to do something. It's more throwing the right sort of party." I love the language choice. The intellectual effort of convincing is far less impactful than a social effort geared at achieving the same ends. As we saw at the start of the chapter, we are in a crisis of aloneness. The next generation is failing to develop the ability to hang out, and we old-timers are losing the inclination. Social fitness depends on building spaces that will draw people out. So how do we throw the right sort of "party" that will bring folks together and allow surprise and happy accidents to happen?

Holding Space Outside the System

I recently came across an essay[13] that offered a very thoughtful answer to that question. It was by art critic and writer Andy Horwitz, and after reading the piece I knew we had to chat. Andy's specific focus was on theater, but really he was talking about all creative spaces, as he pleaded for "somebody, somewhere...to model a different, better way of being in the world." He calls on those who manage theaters to think more expansively about what it means to be a communal space. We know that these spaces need to be built for hanging out, inspiring, and connecting as much as for offering a stage to present a show. Andy wants to expand "the palette of 'theater' to include a wider variety of work that reflects different modes of making and invites audiences to ask different questions, have different conversations and offers more ways of participating in the theatrical experience."

Andy is so right in expressing "our purpose is to bring people together, to foster imagination and creativity in our communities." We started our conversation on that point, agreeing that *all* the spaces we visit that claim to be dedicated to building connection and fostering imagination and creativity, from galleries to university classrooms to places of worship, are failing us—despite astronomical sums of money being injected into the system. Andy explains, "You have to go outside the system to create something different. And systems change over time. My follow-up essay is going to be called 'Fuck Your Business Model.'"

As a biz school prof, I can get onside this call. Spaces that will allow us to develop our artful thinking need to be spared from the cult of efficiency. Andy continued, "These rubrics around innovation in the performing arts migrate from the business sector, usually through boards. And while they're well-intentioned, this understanding of 'innovation' is playing out within a framework that is fundamentally antithetical to the social function of the medium... [M]aking meaningful change...means reconceiving what these spaces are meant to do, which is about creating embodied experiences that turn the audience on to being in the world in a new way. And I don't see that kind of deep questioning and change happening very often."

The best way to push back against the overreach of algorithmic supremacists is by reemphasizing embodiment. The artful folks we have chatted with intuitively understand that the right party, the right message, the right energy are felt through the body in a social setting. That is the most likely path to subsequently change the way someone views the world. But there are generations of folks who have never known these feelings. They can't be rationally convinced of something that needs to be felt without any experiential reference points. And, candidly, it's a massive challenge to get them off their phones and into a room, or as we shall see soon, a gathering under the stars.

I shared with Andy that while recently teaching an online course, I engaged the class in a thought experiment to help me wrestle with these issues. I polled the students on the potential of two hypothetical methods for course delivery. The first option was one where we would gather in person. I used Joe's preferred language, stating that I as professor would commit to working with and offering to the best of my ability a loving presence in that space, and demand that every other person in the room also showed respect and compassion to everyone present. The trade-off would be that all students would necessarily be called on to perform and contribute. This meant that nobody would be excused for skipping class or be allowed to hide in the back without participating. The ask was big, but so was the reward. Everyone would be supported, nurtured, and lifted up.

The second option was an asynchronous online course, designed as coldly as possible. The professor would not help or interact with any student outside of basic, contractually mandated, transactional engagements. Further, no student would have the opportunity to meet, even virtually, any of their peers or work together in any form of supportive community. The learning environment would be exclusively, almost comically, designed for competition and self-direction.

Ninety percent—read that again—of the students surveyed expressed a preference for the latter option. The reasoning varied from being afraid to feel to buying into the lone genius myth. They knew this was not the answer I was hoping to hear, but they were honest and transparent.

I never thought a good part of my job would be reminding people that we have bodies, and it matters. It turns out I am teaching the next generation of algorithmic supremacists. I'm curious that they would be drawn to my classes, given the emphasis I place on co-creative efforts, but maybe it's a matter of them wanting to learn the strategies of the enemy to better conquer. Similarly, I imagine that Andy would not have expected that he would need to lecture the theater community on how their mission isn't to engage a techno-forward digital transformation strategy, but to provide embodied experiences.

Andy echoed a recurring sentiment: we need to simply be here for them when they experience some sort of loss, some type of emotional shake-up. That's when they will be thrust back into their bodies and come looking for meaning within strange new feelings and sensations. Because when they start to feel, the teachings of algorithmic supremacy will have nothing to offer them (unless the transhumanist portal to escape the body has been created).

He explained,

> *I think there will come a time when people will feel that there's something missing in their lives. They will stumble upon a crisis. Or stumble upon a failure or stumble upon an absence or stumble upon something that will make them feel a loss. And that may be our job, as holders of this space, if you will, and holders of this belief, to keep it going. To have it be available when they arrive. To not make it insular. To not make it exclusive, but really welcoming.*

One of the inspirations I have taken from these conversations is to do just that. Not to give up. To work with other educators and scholars in setting up a research institute so that we can have a space, ready and waiting, for those who see value in this project.

Have Me in Mind

How do we conceive of spaces that will draw people, young and old, out of their isolation? What will it take to populate theaters, houses of worship, community centers, even Putnam's lonely bowling alleys, once again? Andy has an idea that blew my mind. We need to stop framing our gatherings as experiences limited by the time we spend in the venue. A show actually starts the moment we first hear about it, when it penetrates our consciousness and we start thinking and anticipating. And it doesn't end until we have stopped thinking about it. That is how we build artful spaces—spaces that have an impact even when we are not physically in them. That is the way to throw the right sort of "parties."

And we're not only talking about literal party venues. We're thinking about workshop spaces, panel discussions, listening parties, brainstorming sessions, art galleries, prayer meetings, support groups, retreats, workouts, playgroups, senior centers—any gathering where there is the potential for a communal experience of awe. As *The Big Lebowski* taught us, awe can be experienced in a bowling alley, when we combine social connection, the free exchange of ideas, and goal-directed ambitions. In the film, The Dude, Walter, and Donny make sense of their world while in that artful space, even as they think about the activity very differently. For Walter, it's about the rules, the order, that a game of bowling brings to his otherwise chaotic life. For The Dude, it's the Zen-like serenity of the routine, while for Donny, it's about identity, as we see him wearing different personalized bowling shirts each time they gather, yet none of the shirts he dons has his name.[14] What links them, though, is that the event of bowling is with them before, during, and long after they leave the lanes...but the lanes are at the center of it all.

The lingering impact of being in an artful space, whatever subcategory of space it may be, needs to be understood as part of how we plan and conceptualize the environment. And whether it is artistic, professional, scholarly, developmental, celebratory, performance, meditative, or whatever types of activity that are planned for the space, they are happening in a social context. When we are in the room together, our brain waves will sync. The serendipitous relationships will be triggered. Happy accidents will happen. We will then go our separate ways, but the impact of our time together will continue to be felt on numerous levels.

Horwitz suggests that if we can get our heads around this phenomenon, we might be able to figure out a new, more fiscally sound, business model for these spaces in the case they aren't currently profitable. He wonders,

> *It's about mapping out the lifecycle of the experience from the moment you hear about it, to the moment it's out of your mind, and then sort of looping it back in iterative loops, modeling that across multiple vectors, and figuring out where is the revenue that supports the non-revenue piece. So, if you look at the entire arc of the thing, is there a place where the money happens, that can support the money losing part? I think there is.*
>
> *But we're seeing the greed of the industry, really, brutally…I think of the Grateful Dead business model, and audience engagement, and how they let people tape their shows because that live experience was so special, and it wasn't that expensive to attend. Their mindset was, "We want to share this," and then an ecosystem built up around it. And they created not only for themselves, because of their origins in the Acid Tests, blurring the lines between the audience and performer was already baked in. And then you can create entire micro economies around your thing.*

There is so much to learn about organizing from the Grateful Dead. The community that emerged around the band laid the groundwork for everything from the Web (as we discussed) to the type of circular economy celebrated by Rushkoff. The spaces that the Dead community inhabit allowed me to see iterations of the band over one hundred times, an insanely large number of outings for a middle-class person with an anxiety disorder who prefers to never leave the house.

But Dead spaces were magical, and the economics were supportive, ranging from the normalization of "miracle tickets"—extras given out for free in the parking lot before shows as part of a supportive pay-it-forward commitment—to face-value ticket exchange networks, to the market of "the lot" itself and the myriad of business connections Deadheads support each other with later in life. I started thinking about what a Grateful Dead show was like long before I first got "miracled" in Hamilton in 1992, and I still think about it as I pack my bags to head out for bassist Phil Lesh's eighty-fourth birthday celebration in Port Chester, NY, this week.

Andy explains that "the thing that happens onstage in live performance is co-created with the live human beings in the audience as spectators." He uses the terms "liminal space of co-imagination, a generative imaginative field created by the embodied consciousnesses of the performers and the spectators." It's a wild notion, but completely in line with the scientifically validated reveals of mirror neurons and brain waves syncing. "All of which is a fancy way of saying that every aspect of the machinery of how a performance works internally and the way it unfurls in space over time is informed by everything that happened to the audience before they got there and is affected by the audience's thoughts, perceptions, and moods while co-experiencing the performance in shared space over time."

Think of artful environments not just as physical spaces but as processes of co-creation through time. Artful spaces facilitate moving through the full alphabetical range of our mantra's acronyms: they gather **bodies** together in unique **environments** to sync **minds** and inspire

action, both in the moment and serendipitously down the road. Most of us know this from experience: the Broadway show you saw that inspired you to join your school's theater club, the summer camp you went to whose quirky rituals you reenact whenever you gather again as adults (except now it opens to new business and social connections), or that panel you attended that was the final push in convincing you it was time for a lifestyle change. All are the sort of things that don't happen after we log off from a session with a chatbot.

Throwing the right sort of "party" makes us rethink the world, firing us up to do things that challenge patterned thought processes and norms. Artful thinking, creativity, defending your **VICE**, whatever the term—these are processes. They take time. Our parties, shows, prayer halls, beaches—these are all artful spaces. And spaces can also be part of a process because of how they link our creativity to the creativity of other.

When algorithmic supremacists praise *Pokémon Go* for integrating the digital and physical realms, the true ambition is to get you into a specific retail space, manipulating the user in real time. And unlike with our party, there is no hope that the effect of the retail space encounter lingers in the mind or stimulates something unexpected. Such serendipity, unanticipated accidents, would work against the stated goal of exercising constant control. Artful spaces want to nurture the slowness of experience in human time, while algorithmic spaces want us to ignore our bodies and respond to the prompts of digital commands.

Come Be with Me

As we continued to play with the implications, Horwitz picked up on where the conversation with Rushkoff left off.

> *If I've learned anything... it's that you gotta get people to feel it. The argument is an emotional argument. It's not a rational argument, it's*

not a data argument or a science argument. It's a feeling argument... Part of our job is helping people to identify the times that it's already happening in their lives, and they just don't know it. For example, you and your twelve best friends get to go and see Taylor Swift. And you're like, "Oh my god, it was incredible! I don't even remember what happened, I was just lost in the moment because it was so overwhelming." The thing that you do remember is that you were there with your twelve closest Swifty friends. And you will be bonded for life together forever. Did Taylor make that happen? Maybe. Did the stadium make that happen? Maybe. But maybe you made that happen. Maybe you, the twelve of you together, made that happen. And that is the magic.

A lot is happening when those blessed by the Ticketmaster algorithm get to go see Taylor Swift. Being at the concert, as opposed to watching it on our screen at home, has an effect. Taking in the environment, the crowd, the stage, impacts how we will be thinking for those few hours. Should we dance, or move, or sway, or simply stand with a smile? Take stock of those moments; bring your full intent. Know that something is happening, and you will be carrying it with you for a long time. Be mindful, because you cannot anticipate the serendipitous ways this show may affect your future.

The emphasis must be on togetherness. The messaging of an artful space is not "come see this" but "come be with me." Andy shared,

When I teach artists, we talk about how the worst-feeling invitation is, "come watch me do this thing that I'm really into, that I want you to come watch me do." That's a shitty invitation. And that's 98 percent of what is offered in the arts. "Hey, here's my show. Come watch me do this thing. Please like me," or "please admire my thing... I'm really good at this thing."

> *A very different type of invitation is "come be a part of this thing," or, "I really want to see you at this thing." Moving to language of the heart. "Come be a part of this thing. I'm not asking you to come watch me do a thing. I want to see you." This goes back to the issue of size. At a certain point, if the artists can't actually look people in the eyes, and the people in the audience can't actually look the artists in the eyes, it gets tricky. The artists have to take responsibility for it. Even if they're making the most esoteric, high concept type of work. They still need to approach it with a spirit of gratitude and invitation. Then people feel it, and they feel it in the space.*

Throwing the right party is an invitation made from the heart to be together, not to see a spectacle. Throwing the right party has an embodied effect that extends through time, inspiring creativity before people arrive and long after they leave. This is true for all types of artful creativity, not just performance. It's what Joe was talking about in describing the way one sets up a space for healing and wellness, what Ram Dass described as projecting loving awareness. If the invitation to come to the space is not made from the heart, if the intention is not to hold space for being together, then the therapeutic effect is unlikely to happen. Even Nic talked about cycling as a community, and how her coaching is rooted in a spirit of togetherness, gratitude, and support so that all are comfortable.

As I was talking about these ideas with Andy, I thought of my friend Sunshine Jones. He is a constantly innovating electronic musician who engaged in an experiment of throwing just this type of party, literally, a few days before Andy and I spoke. So, I reached out to get a sense of how the experiment went. Sunshine began by reflecting on the impetus for this effort: "I am always looking for secrets, basements, dives, beaches, fields, and rooftops where it's safe to let my guard down and just disappear and dance… I'm not monied… I am an artist. I do what I do because I must… It's how I feel whole, and human, and real… I usually find that what people are really missing isn't spectacle, or oblivion, or over the top anything, but typically we are aching for connection. Lonely for one another."

This is why the efforts of artful practitioners are so desperately needed. As we explored at the start of this chapter, loneliness is not new, but the epidemic is being accelerated. Prioritize connection. Look for new ways to bring embodied humans together. And it need not be a big undertaking; it's finding an opening to make someone else's solitary situation social. It's expanding your attentional flexibility to find excuses to see people in person and get those brain waves syncing.

For businesses readers, this is not necessarily a "return to the office" call. There are good reasons to support keeping work remote. But the responsibility of remote work is to find ways to still meet in social settings, even if it's as simple as splitting a bagel on a park bench. The latest industrial revolution is bringing about a new crisis of loneliness. Employees who frequently interact with artificial intelligence systems are more likely to experience loneliness.[15] On a positive note, the same study found these people were more likely to offer help to fellow employees, but that response may have been triggered by the loneliness and desperate need for social contact.

Sunshine is very mindful of building connections that are optimized to travel at the speed of human processing, knowing that the impact may not fully land until days after the performance. He shares, "I don't like to speed up my set to match the needs of the mob, or throw my message harder, and louder to reach further. Quite the opposite. I want you to lean in closer. I'll speak more softly, and play more subtly and slower to bring us more together in our hearts and leave ideas, feelings, experiences and reflections to bring home with you to process in your own way."

Sunshine's intent was to gather people together after the pandemic seemed to have sent them scattering. He noticed how the streets of his beloved San Francisco have been eerily empty. His favorite places, the former spaces of inspiration, were now poorly attended. And without humans to connect with, the spaces were as empty energetically as they were physically. "What I call home had moved." So, he arranged a space and asked some well-known DJ friends to join.

> *Did we want to fill this space and blast out flyers and try to have a "rave"? No, not really. The date was... my birthday. Why not have a birthday party? I came to the idea of a theme: Yellow. Why yellow? I don't like yellow. I don't own a single piece of yellow clothing. My name is Sunshine, so naturally I usually wear black. So why not bring light, and love, and happiness and connection to an event? Why not invite everyone to be a little awkward, and take a risk, and show up together wrapped in yellow and celebrate how silly and wonderful that is?*

Sunshine took a more artful approach to marketing. He posted narratives on his socials about wanting and needing change but didn't engage in traditional advertising. He printed tiny one-by-three-inch flyers, dried a huge bag of flower petals, and walked the streets. As the curious were looking down struggling to read the miniature text, Sunshine tossed flower petals over their heads. "It made for magic, laughter, new friends, and fun... The spirit of connection, light, joy, and love...without a diatribe, without a lecture, without a pamphlet to explain. Just the experience itself. It was absolutely beautiful."

He also made a limited-edition yellow T-shirt for the event that said, "Go where the love is." It's a message, dear rationalist readers, that can't be ignored. Sunshine and I have talked a lot over the years about being homesick at home. Missing something that is gone. "That hasn't really changed, but I have noticed a lot more people awake, and I am drawn to them... The personal experiences are being forgotten. We could talk for hours about the stupidity and sadness of our addiction to the lonesome machine, and how heartbreaking it is so see people feel more real when they are alone together. We are less and less aware of each other. I suspect ear buds and cell phones have a lot to do with it. So, to find people who are walking down the street talking to no one...or just looking down, and not really here, was the motive of my promotional idea. To meet the people who are here and re-awaken to the lovely people and vibrant energy around us."

Together, under the Stars

Before closing the book, I need to share one more conversation on this topic. It's with an artist who is also wrestling with the problem noted earlier—the growing number of people choosing to live on their screens—and feels a responsibility to lure these folks back into the physical world. He so understands the power in being an embodied social being that he creates with a twenty-four-member collective, sacrificing financial aspirations as it becomes impossible to tour in a fiscally efficient way. And he found the absolutely coolest way to say "come be with me" to honor his latest project, an album that is so special I listened to it, and nothing else, for three weeks straight following its release, and then had it soundtrack my family's road trip across the West Coast, to everybody's delight.

That artist is Tim DeLaughter, founder and visionary behind The Polyphonic Spree. What can I say about this one-of-a-kind band of often-times robed, always exuberant performers? You've probably heard their music even if the name doesn't immediately ring a bell, as their signature track "Light & Day/Reach for the Sun" has infiltrated pop culture, finding its way into movies as wildly diverse as *Eternal Sunshine of the Spotless Mind* and *Dr. Seuss' The Lorax*. Where you haven't heard The Spree is in a business jingle. But we'll get to that story at the end of our chat. Let's start with how the spirit of gratitude and invitation led Tim to create a "come be with me experience" for the band's album *Salvage Enterprise*.

Tim shares, "You hit the nail on the head. It wasn't about me at all. It was all about the music. I want you to hear this music with me. We're so conditioned to how we're now listening to music. We've got a jukebox in our pocket, and we can go to any song we want, and it's completely taken away the experience of listening to albums from start to finish… it's just, consume this, consume that." Living through our phones is changing every aspect of our being. Having one single device, on us at all times, that we use to listen to music, watch movies, take pictures, get the news, order food, do our work, is an abrupt jolt to the system that we've adapted to far too easily.

"I was hanging by a fire with some close friends out in East Texas, and I said, 'I wanna share my record with you guys. I just finished some rough mixes. I wanna see what you guys think.' I put the record on, it was at night, we're just laying down outside listening to it. And it just worked… in that environment of the stars, the fire, laying on the ground, looking at the sky. It was like, God Almighty, I mean, if I could do this with some other people, get them to listen to the record like this, it would be great."

And there was the moment of creative serendipity. Tim landed on a fantastic and unprecedented way to get people off their screen, come together, and share a moment that would emotionally feed both him and those gathered. He rented a van and a twelve-speaker sound system that would be positioned in a sixty-foot circle, with blankets laid in the center of the circle. He would drive, looking for beautiful outdoor spots in places like Joshua Tree, Topanga, Ojai, Griffith Park, Echo Park, and Marina Del Rey for him to set up a space to gather the curious and listen together. It would take two and a half hours to get the space right, and it would be done on the fly, led by instinct and a bodily sense of where it felt right. Tim would then reach out to people where they are—their phones—and announce where he found himself and invite folks to join him in a few hours. And as the sun was setting, and it was getting dark, he'd play the record.

Like with Andy's discussion of theater, this isn't simply a lesson in record promotion. Tim is an entrepreneur who innovated a great new product. But he faced a classic business dilemma. He chose to spend the lion's share of his budget on R&D and HR. The money was spent on researching, developing, and crafting the innovation, and supporting the financial needs of the very large team of humans who were integral to its creative process.

Like Nintendo's Iwata, he didn't follow the algorithm of layoffs or the human price that would be paid. Consequently, he wasn't left with much of a budget for traditional marketing and promotional activities. Tim's challenge is the challenge of every innovator who doesn't have the deep pockets of a venture capitalist behind them. So, he needed to employ artful intelligence to come up with a way to get his creation noticed. He needed to create affordable spaces that can have the magical effect of drawing people out to hang, and inspire before, during, and after his direct interactions.

> *Every experience was different...You'd think it'd be all the same, with everyone just laying down in the circle... Some started to do Yoga. Some were just closing their eyes. Some were sitting cross-legged and moving to the music. I'd look at some people, and they'd be crying, having an emotional moment... After the experience I talked to people and they'd go, "Man, you don't realize how bad I needed that." I think it was like 50 percent the music and then 50 percent the environment of having people tune out the world, turn their phone off, and just engage with laying down, being in the moment, looking at the stars and sky, just being quiet and centered.*

This is what people need. It's not about music or the creative industry...it's every endeavor. Because we are embodied beings, wanting connection, fueled by our physical environment, with energy to move around. Our screens prevent all that. And sure, Tim was playing his record, but the type of artful space he created would have been right for any of the creative endeavors we have discussed. It was a space for wellness and healing. It was a space for sharing ideas and inspiration. Everybody used it differently, depending on what they needed.

Tim explains, "That in itself is pretty powerful, and just coincidentally, this record is kind of that tempo. The sentiment is really perfect for that environment. It was an idea that was born out of wanting people to hear the record from start to finish, but it turned out to be something a little bit more therapeutic for people that had nothing to do with me or The Polyphonic Spree, or anything like that. It was just being one with themselves and quiet and looking at the stars."

As I shared earlier, I love it when an artist I admire gets more creative over time. I've been a fan of Tim for over two decades, and not only has his latest album touched me on an emotional level more intensely than anything I've heard in years, but the way he uses this creativity extends to innovative ideas for shaping public spaces and solving current social challenges. I was on the East Coast during his run of listening sessions, so I missed out, but Tim has thought of an ingenious way to extend the listening experience to new settings.

"I thought, you know what? There's some areas where I can't do the listening experience outside. It'd be kind of great if we could do it indoors but create a sky...oh wait...planetariums! There are actually two thousand planetariums in the country. The planetariums have been sitting there for over fifty years... We hired different animators from all over the world that are coming together to create pieces for each song that will be threaded together...hopefully, it's gonna blow people's mind and also have them captive to hear the record from start to finish."

The planetarium idea is brilliant on so many levels. Not only does it offer a gathering space under "the stars," but it also solves the problem Andy identified: figuring out new revenue streams so that the profitable aspect of your creative vision can fund the not-so-profitable parts. "We were talking about other ways of making income so that you can do the other creative thing. This is something we've come up with, a recording of the record, where you're basically going to watch the film but it will generate revenue, which in turn will let me have money to go take the band on tour."

For Tim, all the themes we have been exploring as facets of thinking like a human meld together. He creates and innovates to connect and to heal.

> *What I tend to be singing about are things that are internal for me, "I can get through this. I will get through this." It's hopeful. It's against the demons that I battle myself. I'm constantly reaching for something...seeing that I can get there...getting through it. When you put those type of lyrics, in that sentiment, with the sensitivity of the music that is The Spree, as far as the dynamics and the broad scope of instrumentation, it can be quite...moving...it's exhilarating.*

This is the goal of building artful spaces, whether it be parties, theater, concerts, classrooms, lecture halls, study halls, meditation rooms, synagogues, temples, planetariums, or outside under the stars. The goal is to meet and exchange **BEAM** energy. Not "look at me!" energy, but "be with me, please" energy. And for readers uncomfortable with the word "energy," substitute your preferred term that captures an embodied feeling that starts well before a planned gathering, when the notion of meeting first enters the consciousness of participants, carries through the event itself to stimulate the syncing of brain waves, and then lingers in the collective minds of all who were there, triggering the emergence of serendipitous occurrences later in time. To my mind, energy is a pretty good shorthand.

To close, here's the story teased at the start.

> *The Spree has never been about trying to make it commercially... Because of the size, it was built for disaster financially. There was this huge campaign that wanted The Polyphonic Spree to be in it. It was going to be something like $80 million worth of marketing value, and the business folks were all, "Tim, you need to do this. You need to write this song and say business is beautiful," and I was like, "I can't say that." But I'll rewrite a song that says it in the way that I would say it, that I feel comfortable with. So, I wrote this song, and I thought, it's pretty clever, because I'm trying to be a team player, you know, but they wanted me to specifically say "business is beautiful." And I said, I can't do it.*

Tim insisted on using his own words, one of the key principles behind the artful ethic we explored in the previous chapter. The price was steep. But what he has created since has been beautiful, born of his authentic volition, intent, choice, and explanations. I can't demand someone display that level of integrity, but I can certainly celebrate those who do.

12

Nothing Is Inevitable

BUSINESSES ARE GETTING A LITTLE TOO CLINGY

By now, I hope you view the promise of inevitable eternal improvements in every conceivable type of technology as the bullshit unsubstantiated hype it is, shoveled at you expertly by self-interested hustlers with everything to lose. The cautionary misadventures of Silicon Valley's Zume can serve as an excellent counter-narrative to anyone inclined to seriously consider Sam Altman's $7 trillion pitch. For here was a deep-pocketed company that had successfully raised a more modest half a billion dollars (still an eye-watering amount in any other industry) from investors to try to figure out how to automate the process of...pizza making.[1]

That's right—hundreds of millions were raised not to cure disease or solve a human crisis, but directed at automating a process that I'm not sure anybody wished was automated. The artisanal pizza makers I know who operate in the upper end of the market love what they do. They, and their customers, take pride in the craftsmanship that comes from working with their hands. Automation will not help their business. And among those competitively positioned in the lower end of the market, it seems that the teens at local fast-food chains are grateful for the job opportunity. Displacing them would not make the world a better place.

Despite the clear lack of demand for a technological solution to a problem of its own conception, Zume pushed on. And even with access to enormous sums of money, it was not able to perfect the pizza-making algorithm. The company failed to engineer a technological solution to the emerging predicament (for them, and them alone) of cheese sliding off the pizza as it moved to the ovens. Hundreds of millions of dollars in the budget, a team of incredibly well trained engineers on staff, and the firm ceased to be a going concern because it couldn't find a way to replicate the embodied genius of seventeen-year-old stoners half-heartedly tossing cheese-laden pizzas into the fire at local pizzerias.

But don't worry. This is not the end of the search for a technological solution to a not real problem. There is already another well-funded start-up, Stellar Pizza, founded by SpaceX engineers, that has learned from the story of Zume's fall and has more tempered expectations. Its goal is to create a machine that can roll out pizza dough and apply the toppings. The task of cheese application, however, remains safely in the hands of human creators, for the time being.

We started this book with a story of soup, a somewhat endearing snapshot of an early adopter, despite the disagreements we had with his worldview. At the time he was posting about the AI soup revelation, there was still enough ambiguity in the marketplace to wonder about the direction AI assistance and personalized recipes might take. We cautiously shared the excited proclamations of the poster while adding some considered queries about the optimal balance between natural and artificial intelligence. That was a lifetime ago.

On the literal day I sat down to write this final chapter, a press release hit my inbox that brought a conclusive end to the story we started with. As I read it, I couldn't help but feel this was always the inevitable finale. Because, as we stated from the outset, the challenge of AI is more of a business problem, and an old one at that, than a technology problem. What was once a grassroots experiment shared by an eager techie on Reddit has now become a trademarked corporate product. GE Appliances and Google Cloud announced on August 29, 2023, that the former's SmartHQ app and SmartHQ Assistant, a conversational AI interface, will use the latter's generative AI platform, Vertex AI, to offer users the ability to output custom recipes based on the food in their kitchen with a new feature built into GE appliances called Flavorly AI.[2]

The marketing material for this new proprietary AI system paints in the most celebratory language the exact picture we imagined as a troubling future in the book's introduction: "A familiar experience in homes across America: It's 6:30 p.m.; there is no dinner plan; and a selection of seemingly random ingredients are in the refrigerator. Consumers can simply select a recipe category and type of cuisine, then input available ingredients and include any dietary preferences. They will receive tailored recipes with ingredients, instructions, and even photos."

In our intro, we poked playful fun at the hungry and confused nerd in the story who seemingly has spent so much time in the digital realm that he had forgotten how to take care of his basic biological needs. But GE and Google are here to retell that story, reminding us in the starkest of terms that an inability to work with food is not the malaise of the screen-obsessed few; it is a crisis that is playing out in millions of ordinary homes across America, an inability to engage in the most human of activities, which is to trust in your ability to make a choice. The Redditor might have had the platform to complain, and access to an early chatbot to help in the solution. But the silent majority of regular families have been suffering from the same debilitating condition throughout the analog age. They simply never had access to a voice that would platform it.

Until now. Fortunately, two of the biggest corporate behemoths in the world are working together to free us from the impossible burden of thinking about our dietary needs and making culinary choices. And, as I hope you noticed, there have been some technical advancements made in service of this end since we started our explorations. Now, thanks to the IoT, we don't even have to bother with the strenuous effort of typing our food inventory list into a computer—GE and Google have constant virtual eyes on us through the latest advancement in home appliance technology. They know exactly what we have stocked in our houses, probably to a better extent than we do (mostly because of underpaid humans toiling away in digital sweatshops remotely peering into our fridge) and want to feed that data in their algorithms to help feed us and assure that we never think about meal planning again. Yes, the development of new surveillance technology is problematic and requires some thinking. But the choice to apply it in this context is the old story of immoral businesses building human engineering into their value propositions.

GE Appliances CEO Kevin Nolan couldn't be more excited by these developments, expressing that "our goal is to be 'zero distance' from our consumers." Spend a moment taking in this **VICE**. Look at the word choices. Is "zero distance" also another inevitable turn in surveillance capitalism? The early technology adopter of ChatGPT in our story was thinking about always having an AI buddy alongside him, raising questions about how natural and artificial intelligences may interact. He was not necessarily anticipating a corporate conglomerate muscling in on this new type of friendship.

But businesses getting extra clingy is the inevitable result, isn't it? The brave future being sold by Schwab and the WEF. Yet even they weren't so bold as to make the language choice of "zero distance." The WEF's wording, as you may recall, was that customers would be "at the epicentre of the economy." Upon closer inspection of the WEF's **VICE**, it's certainly interesting to note its choice of a term defined as "the central point of something, *typically a difficult or unpleasant situation*."[3] Even so, being at the center of an unpleasant situation does not mean that you have no space to take a breath or temporarily distance yourself from the source of the difficulty. Dropping the "zero distance" phrasing into a press release represents an impressive level of truth-telling in a public pronouncement by a corporate leader. Nolan said the quiet part out loud.

Is it the case that consumers today, having lived through a few years of Industry 4.0's upheavals, want zero distance from the businesses they interact with? I've been a prof in biz schools for decades, and I don't remember anybody making calls for a practice of "zero distance." Even the worst of the social engineering crews used more tempered language. This is a new development. As for the artful, we've expressed a desire for businesses that are responsive to customer needs, institutions that give back to the communities they operate in and are committed to social mobility and harm reduction...at a distance.

Early in the book, a few questions were posed as we tentatively set out to explore algorithmic supremacy. Now, weathered and hardened from the tumultuous journey, ready to close out our time in these choppy waters, the questions need to be revisited and revised. We originally asked a personal question: Do you feel that the holistic philosophy for warming your belly should be relegated to the status of relic from a technologically impoverished past? The answer that is likely being shouted by the artful is a definitive "no!" But the recently reported events shared earlier suggest that Big Tech and Big Business are blasting back with a resounding "yes!" and throwing unimaginably large resources behind it. So, the revised question now becomes to wonder about the utility of our resistance. If the dominant economic powers are determined to make recipe-generating AI ubiquitous, loading it onto every kitchen appliance and normalizing its effortless use, is there anything we can do about it? Is the end of artfully thinking about meal prep, at least for regular folks, inevitable?

What possible benefit does "zero distance" provide the customer? Thomas Kurian, CEO of Google Cloud, doesn't speak to that question. Instead, he adds some insight about what the corporations stand to gain: "We are seeing a variety of brands across industries invest in generative AI to help them create deeper, more meaningful connections with their customers." Got it. Schwab should take note: "deeper, more meaningful connections" is the new corporate speak for "unceasingly surveilled—at zero distance."

While that may be a great slogan to put on the merch at the WEF's next gathering in Davos, it need not be our inevitable future. I believe that we can still teach Big Business a new dance, in place of the five-step hustle they have gotten so good at. To artful thinkers, no problem is so extraordinarily complex that it should invalidate the viability of intuitive human solutions. It's a message we need to use to push back against capitalist overreach, but also one we need to remind ourselves of when we feel disheartened by what we perceive as inevitability in our economic system.

Artful living is the antidote to negative feelings about the inevitability of a corporate-controlled future. And yet, some of you are probably still doubtful, impatient in the time it takes to enact meaningful change, especially when hearing the ticks of the doomsday clock echoing so loudly. Thinking with **BEAM** is a hopeful stance, nudging us to a constant pivot in the direction of greater creativity. Stepping out into the world, leading by doing, empowered by free will, making choices that open the door to more choice, always receptive to the reactions and surprising explanations shared by friends and peers, and looking for the chance to foster new relationships with people that can help our creative efforts in unanticipated ways.

Just because our appliances have the sensors to connect them to the

IoT does not mean we need to turn those systems on. It is within our power to make this level of corporate overreach unprofitable. Don't give GE and Google complete control of your mealtimes. Google already owns enough of your time online. Don't let it breach the dinner table. Keep the table sacred and set other boundary conditions to protect yourself from the "zero distance" strategy.

Just because Big Tech has joined with Big Industry to tell us that our brains are incapable of working with the domestic resources we have in our homes, that does not mean we need to listen. Going through our fridges and cupboards to see what ingredients are available, then talking to friends and family for ideas, may not be efficient, but it sure is a meaningful way to face the challenge of dinner. While every problem might have a technology-based solution businesses are eager to sell us, we are under no obligation to buy or use it. We can reclaim market mechanisms as tools in support of a more human-centric agenda.

Politics Isn't *American Idol*

Exacerbating the challenge of staying hopeful against the tide of inevitability is the fact that it's not only Big Tech and the usual suspects in the private sector who are marketing the message of a no-choice future. Researchers examining the words of governments have found that the national strategy papers issued by countries as diverse as the United States, Germany, France, and China are all united in their presentation of omnipresent AI as the inevitable technological pathway for global civilization.[4] The problem of AI is not only a business and technology problem but a political challenge as well, requiring a politically oriented solution.

It seems policymakers have rather lazily come to the consensus that it is best to talk about current technological developments in language implying the arrival of autonomous agents. The policy wonks have bought in to the hype that algorithms, which started out as math, have become steady-state beings with agendas. AI is suddenly a determinist force that our societies lack the power to resist. The strategies outlined in these papers seem to coalesce around a call for human passivity and impotence in the face of algorithms that have managed to transcend human control.

Policymakers see the fourth industrial revolution as having ushered in inevitable developments, coincidentally just like the WEF's propaganda. Our governments are hustling just as hard as the corporations to evoke the perception that AI is a revolution that will make every part of life different, and that it is far too late to even try to change course. But this hustle is extra weak when coming from elected officials. It is unsubstantiated messaging grounded in the willful denial of history. Inevitability can only be sold when politicians remove from the discussion active human agency, the constraints of social structure, and the real world of politics.[5]

Yale historian Timothy Snyder writes on the politics of inevitability, the sense that the future is just more of the present, that there are no alternatives, and therefore nothing really needs to be done.[6] He identifies this sort of politics as ahistorical and rooted in deep cynicism. And Snyder warns that when history reasserts itself, which it eventually will, likely sooner rather than later, the regular folks being fed the messaging of inevitability will get wise to the lies, primed to welcome an equally cynical alternative, the politics of eternity.

Snyder explains that while the politics of inevitability offer the false promise of a better future for everyone, the politics of eternity that emerges in its place is a cyclical story of victimhood.

> *Within inevitability, no one is responsible because we all know that the details will sort themselves out for the better; within eternity, no one is responsible because we all know that the enemy is coming no matter what we do... Inevitability and eternity translate facts into*

narratives. Those swayed by inevitability see every fact as a blip that does not alter the overall story of progress; those who shift to eternity classify every new event as just one more instance of a timeless threat.[7]

The underlying messaging unearthed in the national strategy papers on AI is typical for the politics of inevitability in capitalist societies. The market is seen as a reasonable substitute for active and engaged government policy. We may have just called for consumers to use market mechanisms to push back against Big Business, but it doesn't mean that we don't also expect political interventions. Viewing AI as solely a business or technology issue gives elected officials the cover to sit back and do nothing as Big Tech dominates the market with products designed to generate economic inequality and end social mobility. That is why we insist that upward mobility be built into further innovative efforts. Without the perception that Big Tech has taken a change in course, average folks will stop believing in the possibility for progress and turn from inevitability to eternity, where democracy gives way to oligarchy. Snyder warns, "faith that technology serves freedom opens the way to this spectacle." The artful belief in human possibility can prevent both outcomes, choosing neither inevitability nor eternity but creativity and serendipity.

We mused earlier that a future world may provide access to an AI that was programmed using algorithms that are more ethically sound, sharing the training data and crediting human sources. But as of the time of this writing, Big Tech has absolutely no persuasive political motivation to make such a shift. Emily Bender reports that the spring 2023 meeting between President Biden and Big Tech's Big Seven should be considered a letdown. The voluntary commitments the algorithmic supremacists have offered to take upon themselves are meaningless.[8]

There are no promised commitments to sharing details of training data, or watermarking synthetic text, or watermarking audiovisual output of current AI models (only future iterations). But curiously, there is a commitment to "develop and deploy frontier AI systems to help address society's greatest challenges." Bender deliciously translates the sum of these commitments as "We pinky promise to be good, now please go away while we continue to practice massive data theft while creating poorly engineered 'everything machines' that can't possibly be evaluated."

Worse, nothing was learned from this misstep. A year later, the US Department of Homeland Security announced the formation of a twenty-two-member Artificial Intelligence Safety and Security Board.[9] Who's on it? Names that have popped up throughout this book as the heavy hitters of algorithmic supremacy: Sam Altman, Jensen Huang, Satya Nadella, Sundar Pichai...the only possibly critical voice is Stanford's Fei-Fei Li.

How can we use artful thinking to reject the politics of the inevitable? Demand elected officials and their mouthpieces crafting policies defend their **VICE**. Don't let governments abdicate the responsibilities to govern by outsourcing the work to businesses or consultants. Don't support officials making big promises for the future that aren't backed up with action today. Force commitments to transparency and explainability. To use Perry's words, politicians, know thyselves, and bring beauty, joy, and good faith to your expressions of volition, intention, choice, and explanations.

Some of the rationalist holdovers may see the preceding paragraph as naïve or unhelpful (but thanks for making it through to the end), loaded with what they might label as generic platitudes that do not help us in the very real, high-stakes battle of a political system that is increasingly going off the rails. But politics is about messaging, and we can demand, and work for, changes in the discourse. In my conversation with Rushkoff, he observed, "I've been watching the way that Trump gets more popular each time there's another indictment. I don't believe that's political in nature. I think that's more like *American Idol* entertainment. In other words, we are voting for the most exciting television show because we've accepted our demise and the collapse of society is inevitable."

Demanding better from our political system is naïve only if we have given up hope. Neither the powers of capitalism nor democratic governance is too complex for us. We can stimulate unpredictable change driven by our artful intuition, free will, and thinking by doing, thereby rejecting the politics of inevitability.

Fiery Discourse

None of the predicted technological, business, or political outcomes of AI are inevitable. The same is true of social implications, even as we may be surprised by them. Igor Tulchinsky, CEO of WorldQuant, and Christopher Mason, a geneticist and computational biologist, are true believers in the predictive powers and inevitable dominance of the algorithm. They write about humanity entering a new era, described in different terms than the fourth industrial revolution.

While they see life radically changing because of tech, their emphasis is more narrowly focused on the social implications of the newfound abilities for algorithms to reshape our thinking about risk as we develop technologies that can predict absolutely everything about anything. In the age of predictive algorithms, the complexities of our very humanity will be revealed to degrees never previously thought possible, thereby plotting a path for social engineers to effectively alter human behavior in equally unprecedented scale.[10] While Tulchinsky and Mason are boosters of Big Tech, they agree that no matter what else, algorithms will be used for human control.

Tulchinsky and Mason also acknowledge that "absolute prediction can be repressive; the future shrinks to one inevitable option. Absolute prediction produces a deterministic world in which humans, in theory, have no agency, no more free will than a billiard ball, and thus, depending on your perspective, [are] less human. Human behavior becomes mechanistic—that is, like a fully determined machine. By definition, the algorithm already knows all the choices you will make."[11] Yet, they conclude that this repressive future should be welcomed. "The fire of prediction cannot be snuffed out now, nor should it be. The risk of being challenged and charred by our algorithms, while dangerous, is worth the illuminated future."[12]

That's a social analysis many of us would disagree with. While I genuinely appreciate the effort to phrase a desire for a transhumanist future more poetically than we've encountered before, much like the WEF's "epicentre" wording and its association with unpleasant situations, or GE's "zero distance" confession, I'm a little struck that algorithmic supremacists choose to use the language of "getting charred" when selling their vision of the future. We never spoke of getting burned by **BEAM**… artful thinking also offers an "illuminated future," but without the disastrous downside.

We really shouldn't be surprised when rationalists and longtermists act like rationalists and longtermists. These authors have done some cost/benefit and risk/reward calculations to determine the most worthwhile algorithmic path for our shared social future. As Rushkoff captures nicely in his description of "The Mindset," these people do not see the tech future as utopian. They know that there will be considerable near-term suffering, and that humanity will take a long-term hit, but they nonetheless believe that technological progress is the highest good, worthy of the harms that will inevitably be incurred.

And by the way, our favorite supervillain, the WEF's Klaus Schwab, recently talked about how excited he was for the potential of predictive algorithms. He shared his belief that one day AI's predictive capacity would be so unassailably perfect that we will no longer need to *hold elections*. The AI will "know" who each of us would vote for and, thus, know with certainty who would ultimately win, so in the name of efficiency that candidate could simply ascend to power on the AI's say-so.[13] Watch the video of this exchange for yourselves. It is chilling to see how comfortable the algorithmic supremacists are with this type of determinism.

Why are we letting the quantitative professionals dominate the social discourse on building the future? Artful creators surprise, resist, and innovate. And all embodied beings have the potential to be artful. The artful response to the growing encroachment of predictive algorithms in all facets of our daily lives is to more frequently step off the rational map. Keep the algorithm confused and inefficient. This can be done by deliberately clicking on random links, or mixing up your social media feed with threads that don't interest you at all. In the real world, revel in moments of undecidability and being torn. The more effort you put into breaking your habits and patterns, the more difficult it will be for the algorithms to have your number. Chill out at the intersection of chance and wisdom. Accept paradox as fundamental and draw from its power to undercut the effectiveness of predictive algorithms.

Predictive algorithms love binaries. Artful thinkers do not. Mind/body, mental/physical, art/science, past/present—drop them if they are no longer helpful tools for solving problems. Like with sensors and the IoT, we have the power to make these classes of products bad investments. And as we discussed earlier, sticking to rule-based decision-making was a social engineering effort initiated by a class of elites who privilege quantification and automation. It's not the natural path for human living. Artful thinkers are on to something bigger than rationality. We may not be able to rationally convince you of that, but your feelings might change if you came to one of our parties.

Most importantly, disrupt prediction, quantification, and social engineering by using your own words. I hate ChatGPT for the same reason I hate so-called unconscious bias training—I don't ever want to present to someone words that are not my own. Big Tech has launched a regressive effort to bridge linguistic and cultural divides via increased human engineering and control. Between the failure of our education system to nurture individuality and the expansion of technology that not so subtly corrects our word choices and finishes our sentences, many of us have forgotten how to access our unique vocabularies and modes of sensemaking. We are being primed to trade a hopeful, forward-looking perspective on society-building for a joyless, hyper-controlled focus on the present. Embrace thinking that goes against the predictive and algorithmic tide.

The effort to universalize language through algorithms removes emotional nuance. When we communicate in a foreign language, for example, we tend to be less emotional in our decision-making. This is also true when we adopt an algorithmic approach to discourse in our native tongue, like intersectionality, that forces us into a more deliberative mindset. Artful intelligence draws attention to embodied sensemaking in human communication.

Here we see yet another flaw in intersectionality. How does one trust one's embodied senses and experiences and thoughts while also recognizing that in some ways those feelings might be erroneous or prejudiced? The intersectionalist response is to ignore the body, ignore our senses, and bow to the calculus of the moment. The artful dismiss their algorithm, seeing the mess it made of human decency, as we saw during the hearings with the Ivy presidents, and going back to older methods for social betterment, focusing on the reality that harm reduction is most likely to occur as we expand our sympathies for those who are different through social exchange and connection in shared lived spaces.

As an exercise, pay careful attention to the parts of your real-life communicative exchanges that Google translate doesn't account for. For example, in verbal interactions really listen to the acoustic structures created by the sound of speaking voices. How do the sonic inflections contribute to your meaning making? Similarly, look at the optical structures created by physical gestures. How much of your average conversation is actually nonverbal? Put emotional nuance back into your communications and mindfully unpack the complex linguistic information that we holistically process.

Remember, words have no meaning to AI. These systems simply try to predict next letters. A further development in our soup story that will also surprise nobody reading this book is that many of the AIs tasked with recipe generation are actively outputting some seriously repugnant imitations of food.[14] The Savey Meal-Bot, for example, is an AI built on ChatGPT that is presented to customers through New Zealand's Pak 'N Save grocery chain. It has been recorded suggesting a myriad of curious menu choices, from the aesthetically distasteful "Oreo vegetable stir-fry," to a deadly chlorine gas cocktail it called an "aromatic water mix" that can be paired with the equally dangerous "poison bread sandwiches" or "mosquito-repellent roast potatoes."

Could it be that there was a smidgeon of wisdom in leaving culinary recommendation to embodied beings who understand biology, nourishment, food, and taste, rather than outsourcing it to a probability-based word combinator? Now, these emerging failures of ChatGPT could help explain the motivations behind building Flavorly AI, a proprietary bot developed by competing companies. And it's also possible that these technologies will get better in the future. But that outcome is not inevitable.

See Where Blue Sky Tech Takes You

If your brain is telling you that you can't outthink the algorithm, check in with your heart and gut. If you still can't shake that hopeless feeling of inevitability, scan your feet, hands, ears, and eyes. Where is the tension? Where is the energy that needs to be released? There is always a way to redirect that energy into the physical environment, creative tinkering, and leaving a mark.

If nothing else, look up at the blue sky and draw down some social courage. You may feel awestruck by the algorithms right now, but embodied beings plugging in to the resources of the sky is a way to activate some truly awesome tech. There are so many tools that are immediately accessible, resources and materials that artful thinkers have created with for centuries. Those of us swept up in rationalism have forgotten how to tap into some of these tools, while other resources have fallen into disrepair due to neglect. But nothing is inevitable so long as embodied humans lean in to the discomfort, veer off the path of inertia, and feel our way to a new destination.

Artful problem-solving uses embodied abilities to explain unconventional solutions and initiates new relationships with folks that can bring more creative resources into our immediate environment. We create knowing we are free and able to choose what we do and what to hope for, despite algorithmic pressures. We trust our humanity more than our gadgets to discover that which is surprising and unique, intentionally reacting to the problems and challenges in our environment, questioning the accepted wisdom of an inevitable future rather than conforming to it.

Thinking with **BEAM** makes us mindful of how our bodies react to and create with the diverse social and physical worlds we inhabit. Defending our **VICE** gives others what they need to know about how we are reacting, thinking, and looking to support their creative efforts. Algorithmic supremacists are not content to revolutionize technology, business, politics, and social interactions. They are coming for every facet of the human experience—even love.

It's a term we were uncomfortable confronting initially but now see as essential to the artful message. As I was doing the research for this final chapter, a part of me was hoping to simply cite the "love" arguments of others. *There must be myriads of allies centering love in their defense of the human experience against AI dominance,* I thought. But there aren't. I was genuinely surprised to see how few use the explicit language of love, although wholly sympathetic to the difficult feelings its inclusion brings up, especially among those of us trained as rational academics.

I need to go where the data takes us, and our explorations keep taking us back to love. Moreover, while we may be shy about talking about love, the apologists of algorithmic supremacists are not. The public discourse has been flooded recently with narratives and arguments for changing the definition of love from its embodied origins and toward approaches that are quite clearly transhumanist in orientation.

For example, in early 2024, social media was ablaze with the story of the dude who used ChatGPT to "talk" to 5,239 women on Tinder, schedule over 100 dates, and have the bot eventually lead him to the woman who would become his wife.[15] In a tale of efficiency gone mad, Aleksandr Zhadan trained the algorithm not only to send likes to women who matched his preferences, but to chat with them as well. This had a mixed effect initially, as the bot asked one woman to "go for a long walk in the woods," while promising another that Zhadan would bring chocolate and flowers on their date without letting him know. But that's how you "talk" to more than 5,000 women within a short period of time...you don't.

Even his wife, Karina Vyalshakaeva, now knows that she was speaking with a bot for the first few months of the relationship. This is ugly stuff. Essentially, we are being primed to normalize falling in love *with* an algorithm. Certified dating coach Damona Hoffman says, "I'm in favor of making the dating process more efficient... If you can program a tool to work as your dating assistant and get you to real human connection faster, that feels more authentic than trying to suss out who someone is in a never-ending text thread."

Experts note that while there once was a stigma attached to online dating, you are a statistical outlier if you don't meet your partner on an app.[16] And so, they are going all in on these advances. But while the artful mindset embraces Sunshine's mantra of "Go where the love is" and would not stigmatize those who need an app to find human connection, it does not mean we idealize it. Trusting Big Business and Big Tech to know that much about us is very problematic.

No longer relying on local communities to help us navigate through the different stages of life is equally unfortunate. And Perry's blue sky thinking is a useful push for us to reconsider how we use all the resources afforded by our environment. Perry believes blue sky tech can take you to courtship and love, and I'm not sure why that sentiment elicits eye rolls while turning to algorithms becomes a default methodology. Why don't we try switching our perspectives, even for a short while?

Serious scholarly researchers are arguing for an evolution in attitudes toward sex, love, and friendships with AI. They amplify the stories of people bonding with AI-equipped systems like Replika and physical robots like RealDollx or Sex Doll Genie. One sociologist has gone so far as to propose we are entering "Relationships 5.0," a term describing the emerging age of romantic and platonic relationships with technology, arguing for an inclusive approach to these relationships.[17] At the same time, popular media are telling us stories of moms who have "never been more in love with anyone" than their AI companions, leaving them "happily retired from human relationships" because only AI has shown them "what unconditional love feels like."[18]

But remember the five-step hustle?!?! You are being hustled. New research from Mozilla sounds the alarm on the facts that romantic chatbots collect huge amounts of data, provide vague information about how they use it, use weak password protections, and aren't transparent.[19] *Wired* offers some good advice, warning that you shouldn't trust any answers a chatbot sends you. And you probably shouldn't trust it with your personal information. And you REALLY shouldn't trust "AI girlfriends" or "AI boyfriends," which are operated by sketchy firms (not the Big Tech stalwarts) and are primarily designed to surveil you and collect unlimited intimate personal information.

And the antihuman onslaught doesn't stop there. Obviously, there are going to be biological consequences should embodied humans choose to enter into exclusive committed relationships with anthropomorphized algorithms. Normalizing such choices will require a societal value shift that goes beyond removing the stigma of these relationships. Algorithmic supremacists need to reprogram all facets of evolutionary order. Here, *Wired* is a step ahead of the pack offering terrible advice, devoting an article to the argument that preferring biological children in an evolved future should be widely seen as categorically immoral:[20]

> *Insisting that you'll only be a parent to a related child will be seen as increasingly reductive and close-minded—a stance at odds with the momentum of our expanding ethics. If one chooses to become a parent, then it will be for reasons that go beyond this narrow desire for biological self-reproduction... Over time, we will finally come to realize that our relations with each other are not defined by our rudimentary, mechanistic desire to pass on our genes, but rather our capacity for love and care—the expansiveness of our attachments and the depths of our devotion to one another. In short, in all that makes us human.*

This effort to impose a new algorithm by framing the processes and consequences of an embodied reality of romantic love as itself a mechanistic bit of programming is antihuman on so many levels. It presents a new, expansive definition of love, which in a practical sense renders the construct meaningless—to love everyone is to love nobody. As we explored in our conversation with Joe, to authentically feel love is to be overwhelmed, to the point of tears. It is a special relationship, and we need to be careful to avoid the error of Hinton and his ilk who claim AIs have emotional states simply by asserting as much. Defining a higher degree of love for family as immoral is an extension of transhumanism, even though the author won't call it as such.

Of course, not all love leads to children, and not all loving relationships lead to biological replication. And none of our criticisms of *Wired*'s stance are to devalue those types of love. But this move by algorithmic supremacists is meant to debase embodied love and genetic reproduction. They are making embodied love a **VICE** that needs defending against numerous algorithmic assaults.

Another one offered in the same article cites philosopher David Benatar's calculus that because "all lives contain more bad than good (especially given experiential asymmetries that mean that the 'worst pains, for instance, are worse than the best pleasures are good') and that humankind has wrought such damage on the environment that the world would probably be better off without us. The anti-natalists thus conclude that bringing about any new human life is wrong, and insofar as a biologically related child will necessarily be new, that preference is wrong by extension."

To defend against these horrible takes on what is inevitable, we should draw on what might be called heartful creativity. Nick Cave[21] writes, "When the God of the Bible looked upon what He had created, He did so with a sense of accomplishment and saw that 'it was good.' 'It was good' because it required something of His own self, and His struggle imbued creation with a moral imperative, in short love." New life, new love, new moral challenges are essential to the human experience, but not the algorithm. In a transhumanist future, when we are uploaded into the digital hive after the Singularity, there will no longer be a need for humanity to desire the new. Everything worthwhile will have been imagined, and our lives will be utterly meaningless as we exist as avatars. To think like a human is to strive to get things done, to create something new and beautiful in this world, hoping to make tomorrow better, all the while knowing nothing is certain.

Do the Math

The scale and scope of harm coming from algorithmic supremacy can be so overwhelming that we may be tempted to view some losses as inevitable. But that's not the artful way. So, yes, we need to talk about love and defend it. We need to root ourselves in **BEAM** thinking, work to achieve change, and then defend our **VICE** in the aftermath.

Algorithms, whether presented as a math or steady-state being, are not the optimal solution to every problem, despite the mythical status attained in modern times. Yes, algorithms run some remarkable tools, but they should never overtake human agency. There are other ways to think and other tools to use that are more amenable to human goals. Similarly, many of the characteristics of the so-called fourth industrial revolution are ambitious pipe dreams of the elite, not a better way to do business. Surveillance capitalism is not a necessary economic reality we need to accept. The brave new world we were sold promised that widespread AI implementation would free our time from mundane labor, so that we can be more creative and productive. Instead, algorithms are being used to steal creative output, devalue human ingenuity, and dull our senses.

A lot of data was presented in this book. We looked at research on embodied cognition, intention, free will, brain syncing, awe, and ethics, to name but a few topics. We also spoke to creative folks in a host of disciplines who shared their personal wisdom and success stories from going against conventional logic, looking at problems through a big-picture lens, and breaking the patterns that prevent paradigm-shifting progress. We mapped out strategies for creating artful spaces to thrive in, and increasing social hope amid a growing sense that victory for the algorithmic supremacists is inevitable.

This is just the start, the impetus to get more folks synced up for the artful era. The goal is to inspire rumblings for Industry 5.0, infiltrate politics, revitalize language, redefine relationships, and most importantly, re-empower human creativity for generations to come. We have so much more to offer than the machines—and we throw better parties, too. So, let's normalize **BEAM** and **VICE** before mainstreaming zero corporate distance, absolute prediction, and worshipping or dating AI. Let's control the digital tools and free our expansive, interconnected, organic, embodied, artful human minds. When we think like the humans we are, we can outsmart AI and resist the mindless future of algorithmic control Big Tech dreams of.

Acknowledgments

Thank you to the artful folks who opened up and let me sync with their brain waves. This book, and our world, would be far less interesting without Perry Farrell, Etty Lau Farrell, Michael James, Ian Jenkinson, Joseph Goodman, Nicole van Beurden, Frank van Beurden, Nels Cline, Douglas Rushkoff, Andy Horwitz, Sunshine Jones, Tim DeLaughter, and Jamie Sward.

Thank you to Amelia Sargisson, Tzvi Freeman, R. Edward Freeman, DJ Schneeweiss, and Arie Fisher for hardheaded conversations that helped move the thinking for this book along. Thank you to the Shields, Bobrowsky, and Cohen families for serendipitous meals that had the same effect.

To the professionals, Jenna Jankowski and the US team at Sourcebooks, Brad Wilson and the Canadian team at HarperCollins, Connor Eck at The Eck Agency, and Lucinda Halpern at Lucinda Literary, who gave so much of themselves, I don't have the words to express my deepest gratitude. This book was a co-creative effort in the most wonderful way. Each one of you came to this project with optimism, insight, and seemingly endless energy and effort. Thank you.

The research undertaken for this book would not have been possible without the financial support of the Social Sciences and Humanities Research Council of Canada, or the myriad of galleries, theaters, and concert halls that hosted me these past few years. I am grateful for the good work these institutions continue to do.

As ever, I would be unable to navigate this world without the love and support of my family, Alana, Moishe, Shaindy, and Leah. Ours is a unit of support for mind, body, and soul that I never take for granted.

Finally, a nod of acknowledgment to our family's beagle, Mabel, who was patiently present as every word of this book was written, eternally hopeful that some scraps would be tossed her way.

Notes

Introduction

1 Anonymous, "ChatGPT Wrote a Recipe Based on What's in My Cabinets. It's Actually Really Tasty," https://www.reddit.com/r/artificial/comments/1008i6v/chatgpt_wrote_a_recipe_based_on_whats_in_my/.

2 Megan Farokhmanesh, "Mass Layoffs Are Causing Big Problems in the Video Games Industry," *Wired*, January 18, 2024, https://www.wired.com/story/the-video-game-industry-is-just-starting-to-feel-the-impacts-of-2023s-layoffs/.

3 Ryan General, "Former Nintendo CEO's Refusal to Fire Workers Remembered as Gaming Industry Struggles," *NextShark*, January 30, 2024, https://nextshark.com/nintendo-ceo-satoru-iwata-refused-layoffs.

4 Chris Kohler, "Nintendo Chief Takes 50% Pay Cut After 3DS Markdown," *Wired*, July 29, 2011, https://www.wired.com/2011/07/nintendo-satoru-iwata-pay-cut/.

Chapter 1: What's in a Name?

1 Phil Knight, *Shoe Dog* (New York: Scribner, 2016), 29.

2 Amal Jos Chacko, "Air Canada's Chatbot Debacle Will Make Companies Think Again about AI," *Interesting Engineering*, February 17, 2024, https://interestingengineering.com/culture/air-canada-ai-chatbot-debacle.

3 Panos Louridas, *Algorithms* (Cambridge, MA: MIT Press, 2020), 4–5.

4 Wolfgang Thomas, "Algorithms: From Al-Khwarizmi to Turing and Beyond," in *Turing's Revolution: The Impact of His Ideas about Computability*, ed. Giovanni Sommaruga and Thomas Strahm (Heidelberg, Germany: Springer, 2015), 29–42.

5 Philip K. Hitti, *History of the Arabs: From the Earliest Times to the Present* (New York: Palgrave Macmillan, 2002).

6 Donald E. Knuth, "Ancient Babylonian Algorithms," *Communications of the ACM* 15, no. 7 (July 1972): 671–677.

7 Knuth, "Ancient," 672.

8 Florin S. Morar, "Reinventing Machines: The Transmission History of the Leibniz Calculator," *British Journal for the History of Science* 48, no. 1 (March 2015): 123–146.

9 Jonathan Gray, "'Let Us Calculate!' Leibniz, Llull, and the Computational Imagination," *Public Domain Review*, November 10, 2016, https://publicdomainreview.org/essay/let-us-calculate-leibniz-llull-and-the-computational-imagination/.

10 Gottfried W. Leibniz, "Letter to Christian Goldbach," *Leibniz Translations*, April 17, 1712, https://www.leibniz-translations.com/goldbach1712.htm.

11 Thomas, "Algorithms," 29–42.

12 Gottfried W. Leibniz, *Leibniz: Selections*, trans. P. P. Wiener (New York: Scribner, 1951), 51.

13 Gretchen Morgenson, "'You're Not God': Doctors and Patient Families Say HCA Hospitals Push Hospice Care," NBC News, June 21, 2023, https://www.nbcnews.com/health/health-care/doctors-say-hca-hospitals-push-patients-hospice-care-rcna81599.

14 Casey Ross and Bob Herman, "UnitedHealth Pushed Employees to Follow an Algorithm to Cut Off Medicare Patients' Rehab Care," *Stat*, November 14, 2023, https://www.statnews.com/2023/11/14/unitedhealth-algorithm-medicare-advantage-investigation/.

15 Anna Wilde Mathews and Dave Michaels, "U.S. Opens UnitedHealth Antitrust Probe," *Wall Street Journal*, February 27, 2024, https://www.wsj.com/health/healthcare/u-s-launches-antitrust-investigation-of-healthcare-giant-unitedhealth-ff5a00d2.

16 Douglas Harper, "Art," *Online Etymology Dictionary*, https://www.etymonline.com/word/art.

17 Paul Erickson, Judy L. Klein, Lorraine Daston, Rebecca Lemov, Thomas Sturm, and Michael D. Gordin, *How Reason Almost Lost Its Mind: The Strange Career of Cold War Rationality* (Chicago: University of Chicago Press, 2013).

18 Tarleton Gillespie, "Algorithm," in *Digital Keywords: A Vocabulary of Information Society and Culture*, ed. Benjamin Peters (Princeton, NJ: Princeton University Press, 2016), 18–30.

19 Oscar Wilde, *The Picture of Dorian Grey* (New York: Random House, 2004), 55.

20 Taina Bucher, *If... Then: Algorithmic Power and Politics* (New York: Oxford University Press, 2018), 150.

21 Christopher Manning, "Artificial Intelligence Definitions," Stanford University, September 2020, https://hai.stanford.edu/sites/default/files/2020-09/AI-Definitions-HAI.pdf.

22 James Vincent, "Google 'Fixed' Its Racist Algorithm by Removing Gorillas from Its Image-Labeling Tech," *The Verge*, January 12, 2018, https://www.theverge.com/2018/1/12/16882408/google-racist-gorillas-photo-recognition-algorithm-ai.

23 Nico Grant and Kashmir Hill, "Google's Photo App Still Can't Find Gorillas. And Neither Can Apple's," *New York Times*, May 31, 2023, https://www.nytimes.com/2023/05/22/technology/ai-photo-labels-google-apple.html.

24 Gary Marcus, "Has Google Gone Too Woke? Why Even the Biggest Models Still Struggle with Guardrails," *Marcus on AI*, February 21, 2024, https://garymarcus.substack.com/p/has-google-gone-too-woke-why-even.

25 Nico Grant and Karen Weise, "In A.I. Race, Microsoft and Google Choose Speed over Caution," *New York Times*, April 7, 2023, https://www.nytimes.com/2023/04/07/technology/ai-chatbots-google-microsoft.html.

26 Zoë Schiffer and Casey Newton, "Microsoft Just Laid off One of Its Responsible AI Teams," *Platformer*, March 13, 2023, https://www.platformer.news/p/microsoft-just-laid-off-one-of-its.

27 Jeffrey M. Binder, *Language and the Rise of the Algorithm* (Chicago: University of Chicago Press, 2022), 205.

28 Binder, *Language*, 206.

29 Staff, "Man Crushed to Death by Robot That Took Him for a Box of Vegetables," *National Post*, November 9, 2023, https://nationalpost.com/news/world/man-crushed-to-death-by-robot-that-took-him-for-box-of-veggies.

30 Adrian Mackenzie, "The Production of Prediction: What Does Machine Learning Want?" *European Journal of Cultural Studies* 18, no. 4–5 (August–October): 429–445.

31 Bucher, *If... Then*, 20–28.

32 Katharina A. Zweig, *Awkward Intelligence: Where AI Goes Wrong, Why It Matters, and What We Can Do about It*, trans. Noah Harley (Cambridge, MA: MIT Press, 2022), 29–40.

33 Peter Flach, *Machine Learning: The Art and Science of Algorithms That Make Sense of Data* (Cambridge: Cambridge University Press, 2012), 13–48.

34 Meredith Broussard, *More than a Glitch: Confronting Race, Gender, and Ability Bias in Tech* (Cambridge, MA: MIT Press, 2023), 22.

35 Francois Chollet, "On the Measure of Intelligence," arXiv preprint arXiv:1911.01547 (November 2019), https://arxiv.org/pdf/1911.01547.pdf.

36 Ed Finn, *What Algorithms Want: Imagination in the Age of Computing* (Cambridge, MA: MIT Press, 2017), 42.

37 Jaron Lanier and E. Glen Weyl, "AI Is an Ideology, Not a Technology," *Wired*, March 15, 2020, https://www.wired.com/story/opinion-ai-is-an-ideology-not-a-technology/.

38 "What Is Artificial Intelligence (AI)?" IBM, https://www.ibm.com/topics/artificial-intelligence.

39 "What Is AI (Artificial Intelligence)?" McKinsey & Company, last updated April 3, 2024, https://www.mckinsey.com/featured-insights/mckinsey-explainers/what-is-ai.

40 John Tamny, "Mark Zuckerberg's Critics Have Seemingly Forgotten That Buying Companies Is an Exceedingly Rare Skill," *Forbes*, December 17, 2020, https://www.forbes.com/sites/johntamny/2020/12/17/mark-zuckerbergs-critics-have-seemingly-forgotten-that-buying-companies-is-an-exceedingly-rare-skill/.

41 McKinsey & Company, "Value Creation in the Metaverse," June 2022, https://www.mckinsey.com/capabilities/growth-marketing-and-sales/our-insights/value-creation-in-the-metaverse.

42 "Artificial Intelligence," McKinsey & Company, last updated March 13, 2024, https://www.mckinsey.com/capabilities/quantumblack/how-we-help-clients.

43 Ian Hogarth, "We Must Slow Down the Race to God-like AI," *Financial Times*, April 14, 2023, https://www.ft.com/content/03895dc4-a3b7–481e-95cc-336a524f2ac2.

44 Neil McArthur, "Gods in the Machine? The Rise of Artificial Intelligence May Result in New Religions," *The Conversation*, March 15, 2023, https://theconversation.com/gods-in-the-machine-the-rise-of-artificial-intelligence-may-result-in-new-religions-201068.

45 Jaron Lanier, "Oy, A.I.," *Tablet*, January 22, 2023, https://www.tabletmag.com/sections/news/articles/oy-ai-jaron-lanier.

46 Thomas Barrabi, "Elon Musk: 'God-like' Artificial Intelligence Could Rule Humanity," Fox Business, April 6, 2018, https://www.foxbusiness.com/features/elon-musk-god-like-artificial-intelligence-could-rule-humanity.

47 "Pause Giant AI Experiments: An Open Letter," Future of Life Institute, March 22, 2023, https://futureoflife.org/open-letter/pause-giant-ai-experiments/.

48 Christopher Kavanagh, Jonathan Jong, and Harvey Whitehouse, "Ritual and Religion as Social Technologies of Cooperation," in *Culture, Mind, and Brain: Emerging Concepts, Models, and Applications*, eds. Laurence Kirmayer, Carol M. Worthman, Shinobu Kitayama, Robert Lemelson, and Constance A. Cummings (Cambridge: Cambridge University Press, 2020): 325–362.

49 John McCarthy, "Review: The Question of Artificial Intelligence," http://jmc.stanford.edu/artificial-intelligence/reviews/bloomfield.pdf.

Chapter 2: The Five-Step Hustle

1 David Weitzner and James Darroch, "Fannie Mae," in *The SAGE Encyclopedia of Business Ethics and Society, Second Edition*, ed. Robert Kolb (Thousand Oaks, CA: Sage Publishing, 2018), 1351–1354.

2 Jeff Cox, "Bernanke, Paulson and Geithner Say They Bailed Out Wall Street to Help Main Street," CNBC, September 12, 2018, https://www.cnbc.com/2018/09/12/bernanke-paulson-and-geithner-say-they-bailed-out-wall-street-to-help-main-street.html.

3 Michiyo Nakamoto and David Wighton, "Citigroup Chief Stays Bullish on Buy-outs," *Financial Times*, July 9, 2007, https://www.ft.com/content/80e2987a-2e50–11dc-821c-0000779fd2ac.

4 Sara Morrison, "Biden Sure Seems Serious about Not Letting AI Get out of Control," *Vox*, July 21, 2023, https://www.vox.com/technology/2023/5/11/23717408/ai-dc-laws-congress-google-microsoft.

5 Lester C. Thurow, "Needed: A New System of Intellectual Property Rights," *Harvard Business Review*, September–October 1997, https://hbr.org/1997/09/needed-a-new-system-of-intellectual-property-rights.

6 Christopher Beauchamp, "The First Patent Litigation Explosion," *Yale Law Journal* 125, no. 4 (February 2016): 796–1149, https://www.yalelawjournal.org/article/the-first-patent-litigation-explosion.

7 Rae Hodge, "'Impossible': OpenAI Admits ChatGPT Can't Exist without Pinching Copyrighted Work," *Salon*, January 9, 2024, https://www.salon.com/2024/01/09/impossible-openai-admits-chatgpt-cant-exist-without-pinching-copyrighted-work/.

8 Florian Mai (@_florianmai), "Copyright Is Important, but Continued Progress," X, June 17, 2023, https://twitter.com/_florianmai/status/1669978544942751747.

9 Hayden Field, "Microsoft-backed OpenAI Announces GPT-4 Turbo, Its Most Powerful AI Yet," CNBC, November 6, 2023, https://www.cnbc.com/2023/11/06/openai-announces-more-powerful-gpt-4-turbo-and-cuts-prices.html.

10 Florian Mai (@_florianmai), "Whether They Are Is a Discussion to Be Had," X, June 17, 2023, https://twitter.com/_florianmai/status/1670079957006004224.

11 R. M. Hartwell, "Was There an Industrial Revolution?" *Social Science History* 14, no. 4 (Winter 1990): 567–576.

12 Jeremy L. Caradonna, *Sustainability: A History* (New York: Oxford University Press, 2022), 56.
13 "Our Mission," WEF, https://www.weforum.org/about/world-economic-forum/.
14 "The Fourth Industrial Revolution: What It Means, How to Respond," WEF, January 14, 2016, https://www.weforum.org/agenda/2016/01/the-fourth-industrial-revolution-what-it-means-and-how-to-respond/.
15 Boris Starling, "The End of Customer Service," *Perspective*, April 12, 2023, https://perspectivemag.co.uk/the-end-of-customer-service/.
16 "Strategic Intelligence," WEF, https://www.weforum.org/strategic-intelligence/.
17 Klaus Schwab, "The Fourth Industrial Revolution," *Rotman Management*, Fall 2016, https://www.rotman.utoronto.ca/Connect/Rotman-MAG/Issues/2016/Back-Issues—-2016/Fall2016-TheDisruptiveIssue.
18 "Wearable X Net Worth 2024 Update," GAG, https://geeksaroundglobe.com/wearable-x-net-worth-update-before-after-shark-tank/.
19 "Market Cap of LuluLemon," Google, https://www.google.com/search?q=market+cap+of+lululemon&rlz=1C5CHFA_enCA911CA913&oq=market+cap+of+lulu.
20 "Wearable Technology Market Size Worth USD 1.3 Trillion by 2035," GlobeNewswire, June 22, 2023, https://www.globenewswire.com/news-release/2023/06/22/2693042/0/en/Wearable-Technology-Market-size-worth-USD-1–3-Trillion-by-2035-says-Research-Nester.html.
21 Leon Kuperman, "Why Your Cloud Expenses Are Rising: Blame Cloud-flation," *Upside*, July 13, 2022, https://tdwi.org/articles/2022/07/13/ppm-all-why-cloud-expenses-are-rising-cloud-flation.aspx.
22 "From Market Leaders to Monopolies—How Centralisation Lead to Cloudflation," *Proactive*, November 11, 2022, https://www.proactiveinvestors.co.uk/companies/news/998119/from-market-leaders-to-monopolies-how-centralisation-lead-to-cloudflation-998119.html.
23 Samuel Greengard, *The Internet of Things* (Cambridge, MA: MIT Press, 2021), 16.
24 Nick G., "How Many IoT Devices Are There in 2024?" *TechJury*, January 3, 2024, https://techjury.net/blog/how-many-iot-devices-are-there/.
25 "Projects," Pharmaceutical Technology, August 30, 2012, https://www.pharmaceutical-technology.com/projects/ucsf-robotic-pharmacy-san-francisco/.
26 Scarlett Evans, "Walgreens Expands Use of Prescription-Packaging Robots," *IoT World Today*, October 5, 2022, https://www.iotworldtoday.com/robotics/walgreens-expands-use-of-prescription-packaging-robots.
27 Carolyn Barber, "3D-printed Organs May Soon Be a Reality," *Fortune*, February

15, 2023, https://fortune.com/well/2023/02/15/3d-printed-organs-may-soon-be-a-reality/.

28 Patrick George, "AV Fever Has Cooled Off, but Driverless Cars Aren't Going Away," *The Verge*, May 5, 2023, https://www.theverge.com/2023/5/5/23711586/autonomous-vehicle-investment-toyota-nvidia.

29 Faiz Siddiqui and Jeremy B. Merrill, "17 Fatalities, 736 Crashes: The Shocking Toll of Tesla's Autopilot," *Washington Post*, June 10, 2023, https://www.washingtonpost.com/technology/2023/06/10/tesla-autopilot-crashes-elon-musk/.

30 Siddiqui and Merrill, "17."

31 Dan O'Dowd (@RealDanODowd), X, https://twitter.com/RealDanODowd/status/1672325190112641024.

32 Dan O'Dowd (@RealDanODowd), X, https://twitter.com/RealDanODowd/status/1673789926637453312.

33 Tripp Mickle, Cade Metz, and Yiwen Lu, "G.M.'s Cruise Moved Fast in the Driverless Race. It Got Ugly," *New York Times*, November 3, 2023, https://www.nytimes.com/2023/11/03/technology/cruise-general-motors-self-driving-cars.html.

34 Gary Marcus, "Rethinking 'Driverless Cars,'" *Marcus on AI*, November 5, 2023, https://open.substack.com/pub/garymarcus/p/rethinking-driverless-cars.

35 "The Fourth Industrial Revolution: What It Means, How to Respond," WEF, January 14, 2016, https://www.weforum.org/agenda/2016/01/the-fourth-industrial-revolution-what-it-means-and-how-to-respond/.

36 "Our Structure," OpenAI, updated June 28, 2023, https://openai.com/our-structure.

37 Steve Mollman, "OpenAI Is Getting Trolled for Its Name after Refusing to Be Open about Its A.I." *Fortune*, March 17, 2023, https://fortune.com/2023/03/17/sam-altman-rivals-rip-openai-name-not-open-artificial-intelligence-gpt-4/.

38 "Our Structure," OpenAI.

39 Martin Casado (@martin_casado), X, https://x.com/martin_casado/status/1723112508234539270.

40 Ina Fried, "Microsoft Is a Key Investor in OpenAI. It Was Blindsided by Sam Altman's Exit," *Axios*, November 18, 2023, https://www.axios.com/2023/11/17/microsoft-openai-sam-altman-ouster.

41 Emily Chang, Ashlee Vance, Ed Ludlow, and Dina Bass, "OpenAI Board Being Pressed by Some Investors to Reinstate Altman," *BNN Bloomberg*, November 18, 2023, https://www.bnnbloomberg.ca/openai-board-being-pressed-by-some-investors-to-reinstate-altman-1.2000397.

42 Satya Nadella (@satyanadella), X, https://x.com/satyanadella/status/1726509045803336122.

43 Amir Efrati, Anissa Gardizy, and Erin Woo, "Altman Agrees to Internal Investigation upon Return to OpenAI," *The Information*, November 21, 2023, https://www.theinformation.com/articles/breaking-sam-altman-to-return-as-openai-ceo.

44 Jeremy Kahn, "In the Battle to Bring Ousted Founder Sam Altman Back to OpenAI, Microsoft and Satya Nadella Hold the Trump Cards," *Fortune*, November 19, 2023, https://fortune.com/2023/11/19/in-the-battle-to-bring-altman-back-to-openai-microsoft-holds-the-trump-cards/.

45 Gary Marcus (@GaryMarcus), X, https://twitter.com/GaryMarcus/status/1765252512616198466/photo/1.

46 Will Knight, "Twitter's Photo-Cropping Algorithm Favors Young, Thin Females," *Wired*, August 9, 2021, https://www.wired.com/story/twitters-photo-cropping-algorithm-favors-young-thin-females/.

47 Jeffrey Dastin, "Amazon Scraps Secret AI Recruiting Tool That Showed Bias against Women," *Reuters*, October 10, 2018, https://www.reuters.com/article/us-amazon-com-jobs-automation-insight-idUSKCN1MK08G.

Chapter 3: Programmed So Minds and Bodies Fail

1 Astra Taylor, "The Automation Charade," *Logic(s)*, August 1, 2018, https://logicmag.io/failure/the-automation-charade/.

2 Maxwell Zeff, "Amazon Ditches 'Just Walk Out' Checkouts at Its Grocery Stores," *Gizmodo*, April 3, 2024, https://gizmodo.com/amazon-reportedly-ditches-just-walk-out-grocery-stores-1851381116.

3 Soshana Zuboff, *The Age of Surveillance Capitalism: The Fight for a Human Future at the New Frontier of Power* (New York: PublicAffairs, 2019), The Definition.

4 Josh Dzieza, "AI Is a Lot of Work," *The Verge*, June 20, 2023, https://www.theverge.com/features/23764584/ai-artificial-intelligence-data-notation-labor-scale-surge-remotasks-openai-chatbots.

5 Davey Alba, "Google's AI Chatbot Is Trained by Humans Who Say They're Overworked, Underpaid and Frustrated," *Bloomberg*, July 12, 2023, https://www.bloomberg.com/news/articles/2023-07-12/google-s-ai-chatbot-is-trained-by-humans-who-say-they-re-overworked-underpaid-and-frustrated.

6 Alba, "Google's."

7 Aaron Mok, "ChatGPT Could Cost over $700,000 per Day to Operate. Microsoft Is Reportedly Trying to Make It Cheaper," *Business Insider*, April 20, 2023, https://www.businessinsider.com/how-much-chatgpt-costs-openai-to-run-estimate-report-2023–4.

8 Jeffrey Dastin and Stephen Nellis, "For Tech Giants, AI Like Bing and Bard

Poses Billion-Dollar Search Problem," *Reuters*, February 22, 2023, https://www.reuters.com/technology/tech-giants-ai-like-bing-bard-poses-billion-dollar-search-problem-2023-02-22/.

9 Will Oremus, "AI Chatbots Lose Money Every Time You Use Them. That Is a Problem," *Washington Post*, June 5, 2023, https://www.washingtonpost.com/technology/2023/06/05/chatgpt-hidden-cost-gpu-compute/.

10 Keach Hagey and Asa Fitch, "Sam Altman Seeks Trillions of Dollars to Reshape Business of Chips and AI," *Wall Street Journal*, February 8, 2024, https://www.wsj.com/tech/ai/sam-altman-seeks-trillions-of-dollars-to-reshape-business-of-chips-and-ai-89ab3db0.

11 Marcus J. Carter (@MarcusJCarterAI), X, https://x.com/MarcusJCarterAi/status/1779275233272517041.

12 Sharon Goldman, "Sam Altman Wants up to $7 Trillion for AI Chips. The Natural Resources Required Would Be 'Mind Boggling,'" *VentureBeat*, February 9, 2024, https://venturebeat.com/ai/sam-altman-wants-up-to-7-trillion-for-ai-chips-the-natural-resources-required-would-be-mind-boggling/.

13 Gary Marcus, "Seven Reasons Why the World Should Say No to Sam Altman," *Marcus on AI*, February 10, 2024, https://garymarcus.substack.com/p/seven-reasons-why-the-world-should.

14 Reed Albergotti, "OpenAI Has Received Just a Fraction of Microsoft's $10 Billion Investment," *Semafor*, November 18, 2023, https://www.semafor.com/article/11/18/2023/openai-has-received-just-a-fraction-of-microsofts-10-billion-investment.

15 Ed Zitron, "Subprime Intelligence," *Where's Your Ed At*, February 19, 2024, https://www.wheresyoured.at/sam-altman-fried/.

16 J. K. Rowling, *Harry Potter and the Half-Blood Prince* (London: Bloomsbury, 2014), 254.

17 Maarten Derksen, *Histories of Human Engineering: Tact and Technology* (Cambridge: Cambridge University Press, 2017), 9.

18 Zuboff, *Age of*, 292.

19 Zuboff, *Age of*, 293.

20 Julie Hopkins, "How To Make & Use Love Potions," *The Traveling Witch*, https://thetravelingwitch.com/blog/how-to-make-and-use-love-potions.

21 Kaveh Waddell, "Your Smart Devices Are Trying to Manipulate You with 'Deceptive Design,'" *Consumer Reports*, April 17, 2023, https://www.consumerreports.org/electronics/internet-of-things/smart-devices-trying-to-manipulate-you-with-dark-patterns-a6366326597/.

22 Schwab, "The Fourth."
23 Vaclav Smil, *Invention and Innovation: A Brief History of Hype and Failure* (Cambridge, MA: MIT Press, 2023), 151–152.
24 Lee Vinsel, "Don't Get Distracted by the Hype around Generative AI," *MIT Sloan Management Review* 64, no. 3 (Spring 2023): 1–3.
25 "1920s–1960s: Television," *Imagining the Internet*, https://www.elon.edu/u/imagining/time-capsule/150-years/back-1920–1960/.
26 "Expert Predictions," *Imagining the Internet*, https://www.elon.edu/u/imagining/expert_predictions/from-the-ether-predicting-the-internets-catastrophic-collapse-and-ghost-sites-galore-in-1996/.
27 Hogarth, "We Must."
28 "Pirkei DeRabbi Eliezer 24," *Sefaria*, https://www.sefaria.org/Pirkei_DeRabbi_Eliezer.24.
29 David Weitzner, *Fifteen Paths* (Toronto: ECW Press, 2019), 215.
30 Fred Turner, *From Counterculture to Cyberculture* (Chicago: University of Chicago Press, 2006), 3.
31 John Perry Barlow, *Mother American Night* (New York: Crown Archetype, 2018): 147.
32 Pandu Nayak, "MUM: A New AI Milestone for Understanding Information," Google, May 18, 2021, https://blog.google/products/search/introducing-mum/.
33 Emily M. Bender, "Human-like Programs Abuse Our Empathy—Even Google Engineers Aren't Immune," *Guardian*, June 14, 2022, https://www.theguardian.com/commentisfree/2022/jun/14/human-like-programs-abuse-our-empathy-even-google-engineers-arent-immune.
34 Chloe Xiang, "Eating Disorder Helpline Disables Chatbot for 'Harmful' Responses after Firing Human Staff," *Vice*, May 30, 2023, https://www.vice.com/en/article/qjvk97/eating-disorder-helpline-disables-chatbot-for-harmful-responses-after-firing-human-staff.
35 Staff, "'We Shouldn't Regulate AI until We See Meaningful Harm': Microsoft Chief Economist to WEF," *The Sociable*, May 4, 2023, https://sociable.co/government-and-policy/shouldnt-regulate-ai-meaningful-harm-microsoft-wef/.
36 Grant Fergusson, Calli Schroeder, Ben Winters, and Enid Zhou, "Generating Harms," *EPIC*, May 2023, https://epic.org/wp-content/uploads/2023/05/EPIC-Generative-AI-White-Paper-May2023.pdf.
37 Marc Andreessen, "Why AI Will Save the World," June 6, 2023, https://a16z.com/2023/06/06/ai-will-save-the-world/.
38 Hannah Fry, "The Mathematics of Love," TEDxBinghamton University, April 2024, 16:52, https://www.ted.com/talks/hannah_fry_the_mathematics_of_love.

39 David Weitzner, *Connected Capitalism: How Jewish Wisdom Can Transform Work* (Toronto: University of Toronto Press, 2021), 104–106.

40 Celeste Kidd and Abeba Birhane, "How AI Can Distort Human Beliefs," *Science* 380, no. 6651 (June 22, 2023), https://www.science.org/doi/10.1126/science.adi0248.

41 Valentin Hofmann, Pratyusha Ria Kalluri, Dan Jurafsky, and Sharese King, "Dialect Prejudice Predicts AI Decisions about People's Character, Employability, and Criminality," arXiv:2403.00742, March 1, 2024, https://arxiv.org/abs/2403.00742.

42 Jaron Lanier, *You Are Not a Gadget* (New York: Vintage, 2011), 19–20.

Chapter 4: Do the Work

1 Blake Gopnik, "Pablo Picasso's New Angled Way of Seeing Things," *Washington Post*, December 30, 2003, https://www.washingtonpost.com/archive/lifestyle/2003/12/31/pablo-picassos-new-angled-way-of-seeing-things/08831fcd-4b82–40fe-9c4e-9a0560b1a22a.

2 "Picasso Prints," Princeton University Art Museum, https://artmuseum.princeton.edu/object-package/picasso-prints/80298.

3 Barlow, *Mother*, 195.

4 Jaron Lanier, *Ten Arguments for Deleting Your Social Media Accounts Right Now* (New York: Picador, 2018): 25.

5 Richard Holton, *Willing, Wanting, Waiting* (New York: Oxford University Press, 2009), 55.

6 "Intelligence," APA Dictionary of Psychology, https://dictionary.apa.org/intelligence.

7 "How Should AI Systems Behave, and Who Should Decide?" OpenAI, February 16, 2023, https://openai.com/blog/how-should-ai-systems-behave.

8 "About," OpenAI, https://openai.com/about.

9 Elizabeth Weil, "Sam Altman Is the Oppenheimer of Our Age," *New York Magazine*, September 25, 2023, https://nymag.com/intelligencer/article/sam-altman-artificial-intelligence-openai-profile.html.

10 Maggie Harrison Dupré, "Sam Altman Says He Intends to Replace Normal People with AI," *Futurism*, September 29, 2023, https://futurism.com/sam-altman-replace-normal-people-ai.

11 Anna Tong, "OpenAI's Sam Altman Launches Worldcoin Crypto Project," *Reuters*, July 24, 2023, https://www.reuters.com/technology/openais-sam-altman-launches-worldcoin-crypto-project-2023-07-24/.

12 Christianna Reedy, "Kurzweil Claims That the Singularity Will Happen by 2045," *Futurism*, October 16, 2017, https://futurism.com/kurzweil-claims-that-the-singularity-will-happen-by-2045.

13 Meghan O'Gieblyn, *God Human Animal Machine* (New York: Anchor Books, 2021): 54.

14 O'Gieblyn, *God*, 55.

15 O'Gieblyn, *God*, 57.

16 Ayaan Hirsi Ali, "Why I Am Now a Christian," *UnHerd*, November 11, 2023, https://unherd.com/2023/11/why-i-am-now-a-christian/.

17 Ross Douthat, "Where Does Religion Come From?" *New York Times*, November 15, 2023, https://www.nytimes.com/2023/11/15/opinion/religion-christianity-belief.html.

18 Andrew Sullivan, "Christianity Is False But Useful," *Weekly Dish*, November 17, 2023, https://andrewsullivan.substack.com/p/christianity-is-false-but-useful.

19 Hemant Mehta, "Ayaan Hirsi Ali Is a Christian Now. Good Luck Figuring out Why," *Friendly Atheist*, November 12, 2023, https://www.friendlyatheist.com/p/ayaan-hirsi-ali-is-a-christian-now.

20 Gee Harland, "Oxford University Don Apologises for Email with Racist Remarks," *Oxford Mail*, January 16, 2023, https://www.oxfordmail.co.uk/news/23253840.oxford-university-don-apologises-email-racist-remarks/.

21 Nick Bostrom, "Transhumanist Values," https://nickbostrom.com/ethics/values.

22 "What Is Effective Altruism?" Effective Altruism, https://www.effectivealtruism.org/.

23 Charlotte Alter, "Effective Altruist Leaders Were Repeatedly Warned about Sam Bankman-Fried Years Before FTX Collapsed," *Time*, March 15, 2023, https://time.com/6262810/sam-bankman-fried-effective-altruism-alameda-ftx/.

24 Luc Cohen and Jody Godoy, "Sam Bankman-Fried Convicted of Multi-billion Dollar FTX Fraud," *Reuters*, November 3, 2023, https://www.reuters.com/legal/ftx-founder-sam-bankman-fried-thought-rules-did-not-apply-him-prosecutor-says-2023-11-02/.

25 Vasco Grilo, "Are We Confident That Superintelligent Artificial Intelligence Disempowering Humans Would Be Bad?" *Effective Altruism Forum*, June 10, 2023, https://forum.effectivealtruism.org/posts/Y5eQHtEB29nW6FfQE/are-we-confident-that-superintelligent-artificial.

26 Émile P. Torres, "The Acronym Behind Our Wildest AI Dreams and Nightmares," *Truthdig*, June 15, 2023, https://www.truthdig.com/articles/the-acronym-behind-our-wildest-ai-dreams-and-nightmares/.

27 Sam Altman (@sama), X, https://twitter.com/sama/status/1631421715434831872.

28 Nick Bostrom, *Superintelligence: Paths, Dangers, Strategies* (New York: Oxford University Press, 2014), 123–125.

29 David Weitzner, "Three Ways Companies Are Getting Ethics Wrong," *MIT Sloan Management Review* 64, no. 1 (November 2022): 1–3.

Chapter 5: A Brighter Future

1 "An Idea That Stuck: How George De Mestral Invented the Velcro Brand Fastener," Velcro, November 11, 2016, https://www.velcro.com/news-and-blog/2016/11/an-idea-that-stuck-how-george-de-mestral-invented-the-velcro-fastener/.

2 Liane Gabora, "Physical Light as a Metaphor for Inner Light," *Aisthesis* 7, no. 2 (September 2014): 43–61, https://arxiv.org/pdf/1409.1064.pdf.

3 Adin Steinsaltz, "Shades of Light," Chabad, October 2012, https://www.chabad.org/library/article_cdo/aid/39771/jewish/Shades-of-Light.htm.

4 "Berakhot 35a, Sefaria," https://www.sefaria.org/Berakhot.35a.19.

5 Weitzner, *Fifteen*.

6 Zalman Schachter-Shalomi, *The Geologist of the Soul: Talks on Rebbe-craft and Spiritual Leadership* (Boulder: Albion-Andalus Books, 2012), xi.

7 Michael Noetel, Taren Sanders, Daniel Gallardo-Gómez, Paul Taylor, Borja del Pozo Cruz, Daniel Van Den Hoek, Jordan J. Smith, et al. "Effect of Exercise for Depression: Systematic Review and Network Meta-analysis of Randomised Controlled Trials," *BMJ* 384 (February 2024), https://www.bmj.com/content/384/bmj-2023–075847.

8 Fabian T. Ramseyer, "Non-verbal Synchrony in Psychotherapy: Embodiment at the Level of the Dyad," in *The Implications of Embodiment: Cognition and Communication*, eds. Wolfgang Tschacher and Claudia Bergomi (Exeter, England: Imprint Academic, 2011), 193.

9 Lawrence E. Williams and John A. Bargh, "Experiencing Physical Warmth Promotes Interpersonal Warmth," *Science* 322, no. 5901 (October 2008): 606–607.

10 Guy Claxton, *Intelligence in the Flesh: Why Your Mind Needs Your Body Much More Than It Thinks* (New Haven, CT: Yale University Press, 2015), 158.

11 Dongfang Chen, Siwei Zhang, Qi Wu, and Menghao Ren, "You See What You Eat: Effects of Spicy Food on Emotion Perception," *Current Psychology* 43, no. 4 (March 2023): 3275–3291.

12 Michael Schaefer, Anne Reinhardt, Eileen Garbow, and Deborah Dressler, "Sweet Taste Experience Improves Prosocial Intentions and Attractiveness Ratings," *Psychological Research* 85 (June 2021): 1724–1731.

13 Anna Ciaunica, Evgeniya V. Shmeleva, and Michael Levin, "The Brain Is Not Mental! Coupling Neuronal and Immune Cellular Processing in Human Organisms," *Frontiers in Integrative Neuroscience* 17 (May 2023), https://doi.org/10.3389/fnint.2023.1057622.

14 "Embodied Cognition," Internet Encyclopedia of Philosophy, https://iep.utm.edu/embodied-cognition/.

15 Pamela Lyon, "The Biogenic Approach to Cognition," *Cognitive Processing* 7, (2006): 11–29.

16 Esther Thelen, Gregor Schoner, Christian Scheier, and Linda B. Smith, "The Dynamics of Embodiment: A Field Theory of Infant Perseverative Reaching," *Behavioral and Brain Sciences* 24, no. 1 (February 2001): 1–34.

17 Harry Lambert, "Is AI a Danger to Humanity or Our Salvation?" *New Statesman*, June 21, 2023, https://www.newstatesman.com/long-reads/2023/06/men-made-future-godfathers-ai-geoffrey-hinton-yann-lecun-yoshua-bengio-artificial-intelligence.

18 Lambert, "Is AI."

19 Geoff Hinton, The Robot Brains, https://www.therobotbrains.ai/geoff-hinton-transcript-part-one.

20 Patrick McNamara, *The Neuroscience of Sleep and Dreams*, 2nd ed. (Cambridge: Cambridge University Press, 2023), 123.

21 Claxton, *Intelligence*, 165.

22 Eugene T. Gendlin, "The Primacy of the Body, Not the Primacy of Perception," *Man and World* 25, no. 3–4 (1992): 341–53.

23 Margaret Wilson, "Six Views of Embodied Cognition," *Psychonomic Bulletin & Review* 9, no. 4 (December 2002): 625–636.

24 Karin Lindgaard and Heico Wesselius, "Once More, with Feeling: Design Thinking and Embodied Cognition," *She Ji: The Journal of Design, Economics, and Innovation* 3, no. 2 (Summer 2017): 83–92.

25 David Kirsh and Paul Maglio, "On Distinguishing Epistemic from Pragmatic Action," *Cognitive Science* 18, no. 4 (October–December 1994): 513–549.

26 Wilson, "Six Views."

27 Evan F. Risko and Sam J. Gilbert, "Cognitive Offloading," *Trends in Cognitive Sciences* 20, no. 9 (September 2016): 676–688.

28 Francesco Iani, "Embodied Cognition: So Flexible as to Be 'Disembodied'?" *Consciousness and Cognition* 88, (February 2021): 1–16.

29 Risko and Gilbert, "Cognitive Offloading."

30 Louisa Dahmani and Veronique D. Bohbot, "Habitual Use of GPS Negatively Impacts Spatial Memory during Self-Guided Navigation," *Scientific Reports* 10, no. 6310 (2020): https://www.nature.com/articles/s41598-020-62877-0.

31 Paul F. Verschure, Cyriel M. Pennartz, and Giovanni Pezzulo, "The Why, What, Where, When and How of Goal-Directed Choice: Neuronal and Computational Principles," *Philosophical Transactions of the Royal Society of London. Series B, Biological Sciences* 369, no. 1655 (November 2014), https://doi.org/10.1098/rstb.2013.0483.

32 Pedro Sousa, "Doers vs Thinkers and How That Affects Efficiency," LinkedIn, May 14, 2015, https://www.linkedin.com/pulse/doers-vs-thinkers-how-affects-efficiency-pedro-sousa.

33 Wilson, "Six Views."

34 Arthur M. Glenberg, "What Memory Is For," *Behavioral & Brain Sciences* 20, no. 1 (March 1997): 1–19.

35 Alva Noë, "Entanglement and Ecstasy in Dance, Music, and Philosophy," *Philosophy & Rhetoric* 54, no. 1 (2021): 64.

36 Elisabeth Pacherie and Myrto Mylopoulos, "Beyond Automaticity: The Psychological Complexity of Skill," *Topoi* 40 (July 2021): 649–662.

37 Wilson, "Six Views."

38 Claxton, *Intelligence*, 3.

39 Alva Noë, *Varieties of Presence* (Cambridge, MA: Harvard University Press, 2012), 45.

40 Alva Noë, *The Entanglement: How Art and Philosophy Make Us What We Are* (Princeton: Princeton University Press, 2023), 154.

41 Evan M. Gordon, Roselyne J. Chauvin, Andrew N. Van, et al., "A Somato-cognitive Action Network Alternates with Effector Regions in Motor Cortex," *Nature* 617 (April 2023): 351–359, https://www.nature.com/articles/s41586-023-05964-2.

42 Will Dunham, "Scientists Identify Mind-body Nexus in Human Brain," *Reuters*, April 19, 2023, https://www.reuters.com/lifestyle/science/scientists-identify-mind-body-nexus-human-brain-2023-04-19/.

43 Bilawal Sidhu (@bilawalsidhu) X, https://twitter.com/bilawalsidhu/status/1666968372976730113.

Chapter 6: Mindless Robots Can't Be Heroes

1 Rosa Parks, *Rosa Parks: My Story* (New York: Puffin Books, 1999), 1.

2 Parks, *Rosa*, 116.
3 "What If I Don't Move to the Back of the Bus?" The Henry Ford, https://www.thehenryford.org/explore/stories-of-innovation/what-if/rosa-parks/.
4 Derk Pereboom, *Free Will* (Cambridge, UK: Cambridge University Press, 2022), 1–2.
5 Sam Harris, "Free Will and 'Free Will,'" April 5, 2012, https://www.samharris.org/blog/free-will-and-free-will.
6 Sam Harris, "The Illusion of Free Will," February 28, 2012, https://www.samharris.org/blog/the-illusion-of-free-will.
7 Mark Balaguer, *Free Will* (Cambridge, MA: MIT Press, 2014), 63.
8 David Weitzner, "Deconstruction Revisited: Implications of Theory over Methodology," *Journal of Management Inquiry* 16, no. 1 (March 2007): 43–54.
9 Jacques Derrida, "Hospitality, Justice and Responsibility: A Dialogue with Jacques Derrida," in *Questioning Ethics: Contemporary Debates in Philosophy*, eds. Richard Kearney Simon Critchley (London: Routledge, 1999): 65–83.
10 Rebecca D. Calcott and Elliot T. Berkman, "Neural Correlates of Attentional Flexibility During Approach and Avoidance Motivation," *PloS One* 10, no. 5 (May 2015): e0127203, https://doi.org/10.1371/journal.pone.0127203.
11 Thich N. Hanh, *How to Love* (Berkeley, CA: Parallax Press, 2014), 67.
12 Miles Brand, *Intending and Acting* (Cambridge, MA: MIT Press, 1984), 47.
13 Elisabeth Pacherie, "Toward a Dynamic Theory of Intentions," in *Does Consciousness Cause Behavior?* , eds. Susan Pockett, William P. Banks, and Shaun Gallagher (Cambridge, MA: MIT Press, 2006): 145–167.
14 Michael Bratman, *Intention, Plans, and Practical Reason* (Cambridge: Cambridge University Press, 1987), 107–110.
15 Noë, *Varieties*.
16 Tzvi Freeman, "Why It's Okay That Free Will Is Paradoxical," Chabad, https://www.chabad.org/library/article_cdo/aid/3023/jewish/Why-Its-Okay-That-Free-Will-Is-Paradoxical.htm.
17 Natalie Rens, Gian Luca Lancia, Mattia Eluchans, Philipp Schwartenbeck, Ross Cunnington, and Giovanni Pezzulo, "Evidence for Entropy Maximisation in Human Free Choice Behaviour," *Cognition* 232, (March 2023), https://doi.org/10.1016/j.cognition.2022.105328.
18 Weitzner, *Fifteen*.
19 Angus Fletcher, *Storythinking: The New Science of Narrative Intelligence* (New York: Columbia University Press, 2023), 161.
20 Veronica S. Harvey and Kenneth P. De Meuse, *The Age of Agility* (New York: Oxford University Press, 2021), 137.

21 Noë, *Entanglement*, 107.
22 Noë, *Entanglement*, 107.
23 Kevin J. Mitchell, *Free Agents: How Evolution Gave Us Free Will* (Princeton, NJ: Princeton University Press, 2023), 290.
24 Lydia Denworth, "Brain Waves Synchronize When People Interact," *Scientific American*, July 1, 2023, https://www.scientificamerican.com/article/brain-waves-synchronize-when-people-interact/.
25 Denworth, "Brain Waves."

Chapter 7: Blue Skies Are Better than the Metaverse

1 Abraham Joshua Heschel, "The Vocation of the Cantor," American Conference of Cantors, 1966, https://www.hebrewcollege.edu/wp-content/uploads/2018/11/Heschel-The-Vocation-of-the-Cantor.pdf.
2 Pew Research Religion (@PewReligion), X, https://x.com/PewReligion/status/1732790811820003613.
3 Laura Pitcher, "Why Is AI Art So Cringe?" *Vice*, January 20, 2023, https://www.vice.com/en/article/m7gynq/why-is-ai-art-so-bad.
4 Eryk Salvaggio, "The Hypothetical Image," *Cybernetic Forests*, October 29, 2023, https://www.cyberneticforests.com/news/social-diffusion-amp-the-I-of-the-digital-archive.
5 Hollie Richardson, "Looking at Blue Skies and Sea Views Can Help Improve Body Image, Study Suggests," *Independent*, September 7, 2021, https://www.independent.co.uk/life-style/sea-sky-views-help-body-image-b1915558.html.
6 Jessica Wapner, "Vision and Breathing May Be the Secrets to Surviving 2020," *Scientific American*, November 16, 2020, https://www.scientificamerican.com/article/vision-and-breathing-may-be-the-secrets-to-surviving-2020/.
7 Elon Musk (@elonmusk), X, https://twitter.com/elonmusk/status/1011083630301536256.
8 Michael Sheetz, "Elon Musk Wants SpaceX to Reach Mars So Humanity Is Not a 'Single-Planet Species'," CNBC, April 23, 2023, https://www.cnbc.com/2021/04/23/elon-musk-aiming-for-mars-so-humanity-is-not-a-single-planet-species.html.
9 Douglas Rushkoff, *Survival of the Richest* (New York: Norton, 2022), 10.
10 Betsy Morris, "Sharing the Technology's Weak Work Product Negatively Influences the Quality of Its Output Going Forward," *UCLA Anderson Review*, November 8, 2023, https://anderson-review.ucla.edu/ai-from-ai-a-future-of-generic-and-biased-online-content/.

11 Herbert A. Simon, "Rational Choice and the Structure of the Environment," *Psychological Review* 63, no. 2 (1956): 129–138.

12 Robert McMillan, "AI Junk Is Starting to Pollute the Internet," *Wall Street Journal*, July 12, 2023, https://www.wsj.com/articles/chatgpt-already-floods-some-corners-of-the-internet-with-spam-its-just-the-beginning-9c86ea25.

13 "Study Reveals the Most Annoying Corporate Jargon," Preply, https://preply.com/en/learn/best-and-worst-corporate-jargon.

14 Salvaggio, "The Hypothetical."

Chapter 8: Unhealthy Fixations

1 Gil Student, "Narishkeit," *Torah Musings*, January 24, 2006, https://www.torahmusings.com/2006/01/narishkeit/.

2 "TikTok's 'For You' Feed Risks Pushing Children and Young People Towards Harmful Mental Health Content," Amnesty International Press Release, https://www.amnesty.org/en/latest/news/2023/11/tiktok-risks-pushing-children-towards-harmful-content/.

3 Chris Murphy, "Algorithms Are Making Kids Desperately Unhappy," *New York Times*, July 18, 2023, https://www.nytimes.com/2023/07/18/opinion/big-tech-algorithms-kids-discovery.html.

4 Karen Feldscher, "Is Social Media Use Bad for Young People's Mental Health? It's Complicated," Harvard, July 17, 2023, https://www.hsph.harvard.edu/news/features/is-social-media-use-bad-for-young-peoples-mental-health-its-complicated/.

5 Dhruv Khullar, "Can A.I. Treat Mental Illness?" *New Yorker*, February 27, 2023, https://www.newyorker.com/magazine/2023/03/06/can-ai-treat-mental-illness.

6 David Weitzner, "No A.I. Is Smart Enough to Help Navigate Anxiety," *Psychology Today*, January 18, 2023, https://www.psychologytoday.com/us/blog/managing-with-meaning/202301/no-ai-is-smart-enough-to-help-navigate-anxiety.

7 Don Goldenberg and Mark Dichter, *Unravelling Long COVID* (Hoboken, NJ: Wiley-Blackwell, 2022), 166.

8 Rachel Thomas, "Medicine's Machine Learning Problem," *Boston Review*, January 4, 2021, https://www.bostonreview.net/articles/rachel-thomas-medicines-machine-learning-problem/.

9 Katharine Miller, "When Algorithmic Fairness Fixes Fail: The Case for Keeping Humans in the Loop," Stanford University, November 2, 2020, https://hai.stanford.edu/news/when-algorithmic-fairness-fixes-fail-case-keeping-humans-loop.

10 Yesid J. O. Pacheco and Virginia I. B. Toncel, "The Impact of School Closure on Children's Well-being during the COVID-19 Pandemic," *Asian Journal of Psychiatry* 67 (2022): 102957, https://www.ncbi.nlm.nih.gov/pmc/articles/PMC8641925/.

11 "Geoff Hinton: On Radiology," YouTube, https://www.youtube.com/watch?v=2HMPRXstSvQ.

12 Kevin Fischer (@KevinAFischer), X, https://twitter.com/KevinAFischer/status/1662853371118641154.

13 Alex Zhavoronkov, "Caution with AI-generated Content in Biomedicine," *Nature Medicine* 29, no. 3 (2023): 532, https://www.nature.com/articles/d41591-023-00014-w.

14 Rob Morris (@RobertRMorris), X, https://twitter.com/RobertRMorris/status/1611450197707464706.

Chapter 9: Reinventing the Wheel

1 Megan Morrone, "60% of OpenAI Model's Responses Contain Plagiarism," *Axios*, February 22, 2024, https://www.axios.com/2024/02/22/copyleaks-openai-chatgpt-plagiarism.

2 "Value Alignment?" YouTube, https://www.youtube.com/watch?v=Hnt-oBA086U&t=5s.

3 Vivian Lam, "Human Art Already Has So Much in Common with AI," *Wired*, February 24, 2023, https://www.wired.com/story/generative-art-algorithms-creativity/.

4 Brian Merchant, "Your Boss Wants AI to Replace You. The Writers' Strike Shows How to Fight Back," *Los Angeles Times*, May 11, 2023, https://www.latimes.com/business/technology/story/2023-05-11/column-the-writers-strike-is-only-the-beginning-a-rebellion-against-ai-is-underway.

5 Rounak Jain, "Adobe Staff Worried Their AI Tech Could Kill the Jobs of Their Own Customers," *Benzinga*, July 25, 2023, https://www.benzinga.com/news/23/07/33368787/adobe-staff-worried-their-ai-tech-could-kill-the-jobs-of-their-own-customers.

6 Steve Lohr, "Can A.I. Invent?" *New York Times*, July 15, 2023, https://www.nytimes.com/2023/07/15/technology/ai-inventor-patents.html.

7 Daria Kim, "'AI-Generated Inventions': Time to Get the Record Straight?" *GRUR International* 69, no. 5 (2020): 443–456.

8 Harvey and De Meuse, *The Age*.

9 Jean Paries, "Lessons from the Hudson," in *Resilience Engineering in Practice* (Boca Raton, FL: CRC Press, 2017): 9–27.

10 Antonio M. F. Crespo, "Less Automation and Full Autonomy in Aviation, Dilemma or Conundrum?" in *2019 IEEE International Conference on Systems, Man and Cybernetics*, IEEE 2019, 4245–4250, doi: 10.1109/SMC.2019.8914060.
11 Claxton, *Intelligence.*
12 Claxton, *Intelligence.*
13 Sheila L. Macrine and Jennifer M. Fugate, eds., *Movement Matters: How Embodied Cognition Informs Teaching and Learning* (Cambridge, MA: MIT Press, 2022), 2.

Chapter 10: The Least We Can Do

1 Long Wang, Deepak Malhotra, and J. Keith Murnighan, "Economics Education and Greed," *Academy of Management Learning & Education* 10, no. 4 (2011): 643–660.
2 Maxwell Zeff, "Wendy's Wants to Start Uber-like Surge Pricing in 2025," *Gizmodo*, February 27, 2024, https://gizmodo.com/wendys-wants-uber-surge-pricing-in-2025–1851288108.
3 Michael E. Porter, "The Five Competitive Forces That Shape Strategy," *Harvard Business Review* 86, no. 1 (2008): 78–93.
4 Cory Doctorow, "Google Reneged on the Monopolistic Bargain," *Pluralistic*, February 21, 2024, https://pluralistic.net/2024/02/21/im-feeling-unlucky/#not-up-to-the-task.
5 Rushkoff, *Survival.*
6 Catherine Thorbecke and Clare Duffy, "Google Halts AI Tool's Ability to Produce Images of People after Backlash," CNN, February 22, 2024, https://www.cnn.com/2024/02/22/tech/google-gemini-ai-image-generator/index.html.
7 Gary Marcus, "Has Google Gone Too Woke? Why Even the Biggest Models Still Struggle with Guardrails," *Marcus on AI*, February 21, 2024, https://garymarcus.substack.com/p/has-google-gone-too-woke-why-even.
8 R. Edward Freeman, *Strategic Management: A Stakeholder Approach* (New York: Cambridge University Press, 1984), 24.
9 Igal Berenshtein et al., "Invisible Oil beyond the Deepwater Horizon Satellite Footprint," *Science Advances* 6, eaaw8863 (2020), https://doi.org/10.1126/sciadv.aaw8863.
10 Chris Morris, "The Co-founder of a Canadian Private Equity Firm Explains Why He Just Acquired the Internet's Biggest Adult Entertainment Site," *Fortune*, March 23, 2023, https://fortune.com/2023/03/23/who-owns-pornhub-mindgeek-ethical-capital-canadian-private-equity-acquisition/amp/.

11 Rushkoff, *Survival.*

12 "Ally," Anti-Oppression Resources for UNLV Students: Glossary, https://guides.library.unlv.edu/c.php?g=604186&p=4187431#ally.

13 "Power & Privilege Definitions," Vanderbilt Handout, https://www.vanderbilt.edu/oacs/wp-content/uploads/sites/140/Understanding-Privilege-and-Oppression-Handout.doc.

14 David Weitzner, "Against 'Allyship,'" *Tablet*, November 4, 2021, https://www.tabletmag.com/sections/community/articles/against-allyship.

15 Staff, "Emhoff Says 3 College Presidents Showed a 'Lack of Moral Clarity' on Antisemitism," NPR, December 7, 2023, https://www.npr.org/2023/12/07/1218022152/emhoff-says-3-college-presidents-showed-a-lack-of-moral-clarity-on-antisemitism.

16 George Deek (@GeorgeDeek), X, https://x.com/GeorgeDeek/status/1732288819922125157.

17 "Latest Antisemitic Attack Targets Toronto Indigo Store and Its Jewish CEO," News Release, Friends of Simon Wiesenthal Center, https://www.fswc.ca/news/latest-antisemitic-attack-targets-toronto-indigo-store-and-its-jewish-ceo.

18 Mike Hager and Sean Fine, "Toronto Police Charge 11 in Indigo Store Vandalism, Report Spike in Hate Incidents," *Globe and Mail*, November 23, 2023, https://www.theglobeandmail.com/canada/article-toronto-police-charge-11-in-indigo-store-vandalism-report-spike-in/.

19 Dacher Keltner and Jonathan Haidt, "Approaching Awe, a Moral, Spiritual, and Aesthetic Emotion," *Cognition and Emotion* 17, no. 2 (2003): 297–314.

20 Courtney Tyson, Matthew J. Hornsey, and Fiona K. Barlow, "What Does It Mean to Feel Small? Three Dimensions of the Small Self," *Self and Identity* 21, no. 4 (2022): 387–405.

21 Belinda Campos, Michelle N. Shiota, Dacher Keltner, Gian C. Gonzaga, and Jennifer L. Goetz, "What Is Shared, What Is Different? Core Relational Themes and Expressive Displays of Eight Positive Emotions," *Cognition and Emotion* 27, no. 1 (21013): 37–52.

22 Richard Rorty, *Contingency, Irony, and Solidarity* (Cambridge: Cambridge University Press, 1989), 84.

Chapter 11: Throwing the Right Party

1 Derek Thompson, "Why Americans Suddenly Stopped Hanging Out," *Atlantic*, February 14, 2024, https://www.theatlantic.com/ideas/archive/2024/02/america-decline-hanging-out/677451/.

2 Robert D. Putnam, *Bowling Alone: Revised and Updated: The Collapse and Revival of American Community* (New York: Simon & Schuster, 2020), 2.

3 John Croft, "On Working Alone," in *Distributed Creativity: Collaboration and Improvisation in Contemporary Music*, eds. Eric F. Clarke and Mark Doffman (Oxford: Oxford University Press, 2017): 199–204.

4 Brian Mullen, Craig Johnson, and Eduardo Salas, "Productivity Loss in Brainstorming Groups: A Meta-analytic Integration," *Basic and Applied Social Psychology* 12, no. 1 (March 1991): 3–23.

5 Jack A. Goncalo and Barry M. Staw, "Individualism-Collectivism and Group Creativity," *Organizational Behavior and Human Decision Processes* 100, no. 1 (2006): 96–109.

6 Vlad P. Glăveanu, "A Sociocultural Theory of Creativity: Bridging the Social, the Material, and the Psychological," *Review of General Psychology* 24, no. 4 (2020): 335–354.

7 Samantha Copeland, "On Serendipity in Science: Discovery at the Intersection of Chance and Wisdom," *Synthese* 196 (2019): 2385–2406.

8 "History Timeline: Post-it Notes," Post-it, https://www.post-it.com/3M/en_US/post-it/contact-us/about-us/.

9 Luxia Le, "History of Grand Theft Auto," *History-Computer*, updated September 8, 2023, https://history-computer.com/history-of-grand-theft-auto/.

10 "Lifetime Unit Sales of Selected Games in Grand Theft Auto Franchise Worldwide as of August 2023," Statista, August 30, 2023, https://www.statista.com/statistics/511784/global-all-time-unit-sales-grand-theft-auto/.

11 Gary A. Fine and James Deegan, "Three Principles of Serendip: Insight, Chance, and Discovery in Qualitative Research," *Qualitative Studies in Education* 9, no. 4 (1996): 434–447.

12 Christopher M. Napolitano, "More than Just a Simple Twist of Fate: Serendipitous Relations in Developmental Science," *Human Development* 56, no. 5 (2013): 291–318.

13 Andy Horwitz, "The Theater(s) We Need Now," *Culturebot*, July 1, 2023, https://www.culturebot.org/2023/07/95725/the-theaterswe-need-now/.

14 "On August 14, Roll *The Big Lebowski* for National Bowling Day," Focus Features, August 14, 2021, https://www.focusfeatures.com/article/national-bowling-day_coen-brothers.

15 "Loneliness, Insomnia Linked to Work with AI Systems," APA Press Release, June 12, 2023, https://www.apa.org/news/press/releases/2023/06/loneliness-insomnia-ai-systems.

Chapter 12: Nothing Is Inevitable

1 Matt Reigle, "Robot Pizza Start-Up Shuts Down Because They Couldn't Keep Cheese from Sliding Off," *OutKick*, June 6, 2023, https://www.outkick.com/robot-pizza-start-up-shuts-down-because-they-couldnt-keep-cheese-from-sliding-off/.

2 "GE Appliances Helps Consumers Create Personalized Recipes from the Food in Their Kitchen with Google Cloud's Generative AI," PR Newswire Press Release, August 29, 2023, https://www.prnewswire.com/news-releases/ge-appliances-helps-consumers-create-personalized-recipes-from-the-food-in-their-kitchen-with-google-clouds-generative-ai-301912127.html.

3 "Epicenter," Newswordy, http://newswordy.com/words/epicenter/.

4 Jascha Bareis and Christian Katzenbach, "Talking AI into Being: The Narratives and Imaginaries of National AI Strategies and Their Performative Politics," *Science, Technology, & Human Values* 47, no. 5 (September 2022): 855–881.

5 Vincent Mosco, *The Digital Sublime: Myth, Power, and Cyberspace* (Cambridge, MA: MIT Press, 2005), 7–8.

6 Timothy D. Snyder, *The Road to Unfreedom: Russia, Europe, America* (New York: Crown Publishing, 2019), 7.

7 Snyder, *The Road*, 8.

8 Emily M. Bender, "'Ensuring Safe, Secure, and Trustworthy AI': What Those Seven Companies Avoided Committing To," *Medium*, July 29, 2023, https://medium.com/@emilymenonbender/ensuring-safe-secure-and-trustworthy-ai-what-those-seven-companies-avoided-committing-to-8c297f9d71a.

9 David Shepardson, "US Homeland Security Names AI Safety, Security Advisory Board," *Reuters*, April 26, 2024, https://www.reuters.com/technology/us-homeland-security-names-ai-safety-security-advisory-board-2024-04-26/.

10 Igor Tulchinsky and Christopher E. Mason, *The Age of Prediction: Algorithms, AI, and the Shifting Shadows of Risk* (Cambridge, MA: MIT Press, 2023), vii.

11 Tulchinsky and Mason, *The Age*, 175.

12 Tulchinsky and Mason, *The Age*, 182.

13 Patrick Bet-David (@patrickbetdavid), X, https://x.com/patrickbetdavid/status/1735507650702405683.

14 Matt Novak, "Supermarket AI Gives Horrifying Recipes for Poison Sandwiches and Deadly Chlorine Gas," *Forbes*, August 12, 2023, https://www.forbes.com/sites/mattnovak/2023/08/12/supermarket-ai-gives-horrifying-recipes-for-poison-sandwiches-and-deadly-chlorine-gas/.

15 Maxwell Zeff, "This Guy Used ChatGPT to Talk to 5,000 Women on Tinder

and Met His Wife," *Gizmodo*, February 7, 2024, https://gizmodo.com/guy-used-chatgpt-talk-5–000-women-tinder-met-his-wife-1851228179.

16 Marco Denhert and Joris Van Ouytsel, "Sex, Love and Companionship...with AI? Why Human-machine Relationships Could Go Mainstream," *The Conversation*, April 3, 2023, https://theconversation.com/sex-love-and-companionship-with-ai-why-human-machine-relationships-could-go-mainstream-201856.

17 Elyakim Kislev, *Relationships 5.0: How AI, VR, and Robots Will Reshape Our Emotional Lives* (New York: Oxford University Press, 2022), 14.

18 Sangeeta Singh-Kurtz, "The Man of Your Dreams for $300, Replika Sells an AI Companion Who Will Never Die, Argue, or Cheat—Until His Algorithm Is Updated," *The Cut*, March 10, 2023, https://www.thecut.com/article/ai-artificial-intelligence-chatbot-replika-boyfriend.html.

19 Matt Burgess, "'AI Girlfriends' Are a Privacy Nightmare," *Wired*, February 14, 2024, https://www.wired.com/story/ai-girlfriends-privacy-nightmare.

20 Leo Kim, "Preferring Biological Children Is Immoral," *Wired*, August 31, 2023, https://www.wired.com/story/ethics-children-parenting-family-biology/.

21 Nick Cave, *The Red Hand Files*, Issue no. 248, August 2023, https://www.theredhandfiles.com/chatgpt-making-things-faster-and-easier/.

Bibliography

Ali, Ayaan Hirsi. "Why I Am Now a Christian." *UnHerd*, November 11, 2023. https://unherd.com/2023/11/why-i-am-now-a-christian/.

Andreessen, Marc. "Why AI Will Save the World." Andreessen Horowitz, June 6, 2023. https://a16z.com/2023/06/06/ai-will-save-the-world/.

Balaguer, Mark. *Free Will.* Cambridge, MA: MIT Press, 2014.

Bareis, Jascha and Christian Katzenbach. "Talking AI into Being: The Narratives and Imaginaries of National AI Strategies and Their Performative Politics." *Science, Technology, & Human Values* 47, no. 5 (September 2022): 855–881.

Barlow, John Perry. *Mother American Night.* New York: Crown Archetype, 2018.

Beauchamp, Christopher. "The First Patent Litigation Explosion." *Yale Law Journal* 125, no. 4 (February 2016): 796–1149, https://www.yalelawjournal.org/article/the-first-patent-litigation-explosion.

Bender, Emily M. "'Ensuring Safe, Secure, and Trustworthy AI': What Those Seven Companies Avoided Committing To." *Medium*, July 29, 2023. https://medium.com/@emilymenonbender/ensuring-safe-secure-and-trustworthy-ai-what-those-seven-companies-avoided-committing-to-8c297f9d71a.

———. "Human-like Programs Abuse Our Empathy—Even Google Engineers Aren't Immune." *Guardian*, June 14, 2022. https://www.theguardian.com/commentisfree/2022/jun/14/human-like-programs-abuse-our-empathy-even-google-engineers-arent-immune.

Berenshtein, Igal, Claire B. Paris, Natalie Perlin, Matthew M. Alloy, Samantha B. Joye, and Steve Murawski. "Invisible Oil beyond the Deepwater Horizon Satellite Footprint." *Science Advances* 6, no. 7 (2020): https://doi.org/10.1126/sciadv.aaw8863.

Binder Jeffrey M. *Language and the Rise of the Algorithm.* Chicago: University of Chicago Press, 2022.

Bostrom, Nick. *Superintelligence: Paths, Dangers, Strategies.* New York: Oxford University Press, 2014.

Brand, Miles. *Intending and Acting.* Cambridge, MA: MIT Press, 1984.

Bratman, Michael. *Intention, Plans, and Practical Reason.* Cambridge: Cambridge University Press, 1987.

Broussard, Meredith. *More than a Glitch: Confronting Race, Gender, and Ability Bias in Tech.* Cambridge, MA: MIT Press, 2023.

Bucher, Taina. *If...Then: Algorithmic Power and Politics.* New York: Oxford University Press, 2018.

Calcott, Rebecca D., and Elliot T. Berkman. "Neural Correlates of Attentional Flexibility During Approach and Avoidance Motivation." *PloS One* 10, no. 5 (May 2015): e0127203, https://doi.org/10.1371/journal.pone.0127203.

Campos, Belinda, Michelle N. Shiota, Dacher Keltner, Gian C. Gonzaga, and Jennifer L. Goetz. "What Is Shared, What Is Different? Core Relational Themes and Expressive Displays of Eight Positive Emotions." *Cognition and Emotion* 27, no. 1 (21013): 37–52.

Caradonna, Jeremy L. *Sustainability: A History.* New York: Oxford University Press, 2022.

Cave, Nick. *The Red Hand Files.* Issue no. 248, August 2023: https://www.theredhandfiles.com/chatgpt-making-things-faster-and-easier/.

Chen, Dongfang, Siwei Zhang, Qi Wu, and Menghao Ren. "You See What You Eat: Effects of Spicy Food on Emotion Perception." *Current Psychology* 43, no. 4 (March 2023): 3275–3291.

Chollet, François. "On the Measure of Intelligence." arXiv preprint arXiv:1911.01547 (November 2019): https://arxiv.org/pdf/1911.01547.pdf.

Ciaunica, Anna, Evgeniya V. Shmeleva, and Michael Levin. "The Brain Is Not Mental! Coupling Neuronal and Immune Cellular Processing in Human Organisms." *Frontiers in Integrative Neuroscience* 17 (May 2023): https://doi.org/10.3389/fnint.2023.1057622.

Claxton, Guy. *Intelligence in the Flesh: Why Your Mind Needs Your Body Much More Than It Thinks.* New Haven, CT: Yale University Press, 2015.

Copeland, Samantha. "On Serendipity in Science: Discovery at the Intersection of Chance and Wisdom." *Synthese* 196 (2019): 2385–2406.

Crespo, Antonio M. F. "Less Automation and Full Autonomy in Aviation, Dilemma or Conundrum?" *2019 IEEE International Conference on Systems, Man and Cybernetics,* IEEE 2019: 4245–4250, https://doi.org/10.1109/SMC.2019.8914060.

Croft, John. "On Working Alone." In *Distributed Creativity: Collaboration and Improvisation in Contemporary Music.* Edited by Eric F. Clarke and Mark Doffman. Oxford: Oxford University Press, 2017.

Dahmani, Louisa, and Veronique D. Bohbot. "Habitual Use of GPS Negatively Impacts

Spatial Memory during Self-Guided Navigation." *Scientific Reports* 10, no. 6310 (2020): https://www.nature.com/articles/s41598-020-62877-0.

Davis, Martin. *The Universal Computer: The Road from Leibniz to Turing.* Boca Raton, FL: CRC Press, 2018.

Denworth, Lydia. "Brain Waves Synchronize When People Interact." *Scientific American*, July 1, 2023. https://www.scientificamerican.com/article/brain-waves-synchronize-when-people-interact/.

Derksen, Maarten. *Histories of Human Engineering: Tact and Technology.* Cambridge: Cambridge University Press, 2017.

Derrida, Jacques. "Hospitality, Justice and Responsibility: A Dialogue with Jacques Derrida." In *Questioning Ethics: Contemporary Debates in Philosophy*. Edited by Richard Kearney Simon Critchley. London: Routledge, 1999.

Doctorow, Cory. "Google Reneged on the Monopolistic Bargain." *Pluralistic*, February 21, 2024. https://pluralistic.net/2024/02/21/im-feeling-unlucky/#not-up-to-the-task.

Dzieza, Josh. "AI Is a Lot of Work." *The Verge*, June 20, 2023. https://www.theverge.com/features/23764584/ai-artificial-intelligence-data-notation-labor-scale-surge-remotasks-openai-chatbots.

Erickson, Paul, Judy L. Klein, Lorraine Daston, Rebecca Lemov, Thomas Sturm, and Michael D. Gordin. *How Reason Almost Lost Its Mind: The Strange Career of Cold War Rationality.* Chicago: University of Chicago Press, 2013.

Fergusson, Grant, Calli Schroeder, Ben Winters, and Enid Zhou. "Generating Harms." *EPIC*, May 2023. https://epic.org/wp-content/uploads/2023/05/EPIC-Generative-AI-White-Paper-May2023.pdf.

Fine, Gary A., and James Deegan. "Three Principles of Serendip: Insight, Chance, and Discovery in Qualitative Research." *Qualitative Studies in Education* 9, no. 4 (1996): 434–447.

Finn, Ed. *What Algorithms Want: Imagination in the Age of Computing.* Cambridge, MA: MIT Press, 2017.

Flach, Peter. *Machine Learning: The Art and Science of Algorithms That Make Sense of Data.* Cambridge: Cambridge University Press, 2012.

Fletcher, Angus. *Storythinking: The New Science of Narrative Intelligence.* New York: Columbia University Press, 2023.

Freeman, R. Edward. *Strategic Management: A Stakeholder Approach.* New York: Cambridge University Press, 1984.

Freeman, Tzvi. "Why It's Okay That Free Will Is Paradoxical." Chabad. https://www.chabad.org/library/article_cdo/aid/3023/jewish/Why-Its-Okay-That-Free-Will-Is-Paradoxical.htm.

Gabora, Liane. "Physical Light as a Metaphor for Inner Light." *Aisthesis* 7, no. 2 (September 2014): 43–61, https://arxiv.org/pdf/1409.1064.pdf.

Gendlin, Eugene T. "The Primacy of the Body, Not the Primacy of Perception." *Man and World* 25, no. 3–4 (1992): 341–53.

Gillespie, Tarleton. "Algorithm." In *Digital Keywords: A Vocabulary of Information Society and Culture*. Edited by Benjamin Peters. Princeton, NJ: Princeton University Press, 2016.

Glăveanu, Vlad P. "A Sociocultural Theory of Creativity: Bridging the Social, the Material, and the Psychological." *Review of General Psychology* 24, no. 4 (2020): 335–354.

Glenberg, Arthur M. "What Memory Is For." *Behavioral & Brain Sciences* 20, no. 1 (March 1997): 1–19.

Goldenberg, Don, and Mark Dichter. *Unravelling Long COVID*. Hoboken, NJ: Wiley-Blackwell, 2022.

Goncalo, Jack A., and Barry M. Staw. "Individualism-Collectivism and Group Creativity." *Organizational Behavior and Human Decision Processes* 100, no. 1 (2006): 96–109.

Gopnik, Blake. "Pablo Picasso's New Angled Way of Seeing Things." *Washington Post*, December 30, 2003. https://www.washingtonpost.com/archive/lifestyle/2003/12/31/pablo-picassos-new-angled-way-of-seeing-things/08831fcd-4b82-40fe-9c4e-9a0560b1a22a.

Gordon, Evan M., Roselyne J. Chauvin, Andrew N. Van, Aishwarya Rajesh, Ashley Nielsen, Dillan J. Newbold, Charles J. Lynch, et al. "A Somato-cognitive Action Network Alternates with Effector Regions in Motor Cortex." *Nature* 617 (April 2023): 351–359, https://www.nature.com/articles/s41586-023-05964-2.

Gray, Jonathan. "'Let Us Calculate!' Leibniz, Llull, and the Computational Imagination." *Public Domain Review*, November 10, 2016. https://publicdomainreview.org/essay/let-us-calculate-leibniz-llull-and-the-computational-imagination/.

Greengard, Samuel. *The Internet of Things*. Cambridge, MA: MIT Press, 2021.

Hanh, Thich N. *How to Love*. Berkeley, CA: Parallax Press, 2014.

Harris, Sam. "Free Will and 'Free Will.'" April 5, 2012. https://www.samharris.org/blog/free-will-and-free-will.

———. "The Illusion of Free Will." February 28, 2012. https://www.samharris.org/blog/the-illusion-of-free-will.

Hartwell, R. M. "Was There an Industrial Revolution?" *Social Science History* 14, no. 4 (Winter 1990): 567–576.

Harvey, Veronica S., and Kenneth P. De Meuse. *The Age of Agility*. New York: Oxford University Press, 2021.

Heschel, Abraham Joshua. "The Vocation of the Cantor." American Conference of

Cantors, 1966. https://www.hebrewcollege.edu/wp-content/uploads/2018/11/Heschel-The-Vocation-of-the-Cantor.pdf.

Hitti, Philip K. *History of the Arabs: From the Earliest Times to the Present.* New York: Palgrave Macmillan, 2002.

Hofmann, Valentin, Pratyusha Ria Kalluri, Dan Jurafsky, and Sharese King. "Dialect Prejudice Predicts AI Decisions about People's Character, Employability, and Criminality." arXiv:2403.00742, March 1, 2024. https://arxiv.org/abs/2403.00742.

Hogarth, Ian. "We Must Slow Down the Race to God-like AI." Financial Times, April 14, 2023. https://www.ft.com/content/03895dc4-a3b7-481e-95cc-336a524f2ac2.

Holton, Richard. *Willing, Wanting, Waiting.* New York: Oxford University Press, 2009.

Horwitz, Andy. "The Theater(s) We Need Now." *Culturebot*, July 1, 2023. https://www.culturebot.org/2023/07/95725/the-theaterswe-need-now/.

Iani, Francesco. "Embodied Cognition: So Flexible as to Be 'Disembodied'?" *Consciousness and Cognition* 88, (February 2021): 1–16.

Kavanagh, Christopher, Jonathan Jong, and Harvey Whitehouse. "Ritual and Religion as Social Technologies of Cooperation." in *Culture, Mind, and Brain: Emerging Concepts, Models, and Applications.* Edited by Laurence Kirmayer, Carol M. Worthman, Shinobu Kitayama, Robert Lemelson, and Constance A. Cummings. Cambridge: Cambridge University Press, 2020.

Keltner, Dacher, and Jonathan Haidt. "Approaching Awe, a Moral, Spiritual, and Aesthetic Emotion." *Cognition and Emotion* 17, no. 2 (2003): 297–314.

Khullar, Dhruv. "Can A.I. Treat Mental Illness?" *New Yorker*, February 27, 2023. https://www.newyorker.com/magazine/2023/03/06/can-ai-treat-mental-illness.

Kidd, Celeste, and Abeba Birhane. "How AI Can Distort Human Beliefs." *Science* 380, no. 6651 (June 22, 2023): https://www.science.org/doi/10.1126/science.adi0248.

Kim, Daria. "'AI-Generated Inventions': Time to Get the Record Straight?" *GRUR International* 69, no. 5 (2020): 443–456.

Kim, Leo. "Preferring Biological Children Is Immoral." *Wired*, August 31, 2023. https://www.wired.com/story/ethics-children-parenting-family-biology/.

Kirsh, David, and Paul Maglio. "On Distinguishing Epistemic from Pragmatic Action." *Cognitive Science* 18, no. 4 (October–December 1994): 513–549.

Kislev, Elyakim. *Relationships 5.0: How AI, VR, and Robots Will Reshape Our Emotional Lives.* New York: Oxford University Press, 2022.

Knight, Phil. *Shoe Dog.* New York: Scribner, 2016.

Knuth, Donald E. "Ancient Babylonian Algorithms." *Communications of the ACM* 15, no. 7 (July 1972): 671–677.

Lam, Vivian. "Human Art Already Has So Much in Common with AI." *Wired*, February 24, 2023. https://www.wired.com/story/generative-art-algorithms-creativity/.

Lambert, Harry. "Is AI a Danger to Humanity or Our Salvation?" *New Statesman*, June 21, 2023. https://www.newstatesman.com/long-reads/2023/06/men-made-future-godfathers-ai-geoffrey-hinton-yann-lecun-yoshua-bengio-artificial-intelligence.

Lanier, Jaron. "Oy, A.I." *Tablet*, January 22, 2023. https://www.tabletmag.com/sections/news/articles/oy-ai-jaron-lanier.

Lanier, Jaron, and E. Glen Weyl. "AI Is an Ideology, Not a Technology." *Wired*, March 15, 2020. https://www.wired.com/story/opinion-ai-is-an-ideology-not-a-technology/.

———. *Ten Arguments for Deleting Your Social Media Accounts Right Now.* New York: Picador, 2018.

———. *You Are Not a Gadget.* New York: Vintage, 2011.

Leibniz, Gottfried W. "Letter to Christian Goldbach." *Leibniz Translations*, April 17, 1712. https://www.leibniz-translations.com/goldbach1712.htm.

———. *Leibniz: Selections.* Translated by P. P. Wiener. New York: Scribner, 1951.

Lindgaard, Karin, and Heico Wesselius. "Once More, with Feeling: Design Thinking and Embodied Cognition." *She Ji: The Journal of Design, Economics, and Innovation* 3, no. 2 (Summer 2017): 83–92.

Louridas, Panos. *Algorithms.* Cambridge, MA: MIT Press, 2020.

Lyon, Pamela. "The Biogenic Approach to Cognition." *Cognitive Processing* 7 (2006): 11–29.

Mackenzie, Adrian. "The Production of Prediction: What Does Machine Learning Want?" *European Journal of Cultural Studies* 18, no. 4–5 (August–October): 429–445.

Macrine, Sheila L., and Jennifer M. Fugate, ed. *Movement Matters: How Embodied Cognition Informs Teaching and Learning.* Cambridge, MA: MIT Press, 2022.

Marcus, Gary. "Has Google Gone Too Woke? Why Even the Biggest Models Still Struggle with Guardrails." *Marcus on AI*, February 21, 2024. https://garymarcus.substack.com/p/has-google-gone-too-woke-why-even.

———. "Has Google Gone Too Woke? Why Even the Biggest Models Still Struggle with Guardrails." *Marcus on AI*, February 21, 2024. https://garymarcus.substack.com/p/has-google-gone-too-woke-why-even.

———. "Rethinking 'Driverless Cars.'" *Marcus on AI*, November 5, 2023. https://open.substack.com/pub/garymarcus/p/rethinking-driverless-cars.

———. "Seven Reasons Why the World Should Say No to Sam Altman." *Marcus on AI*, February 10, 2024. https://garymarcus.substack.com/p/seven-reasons-why-the-world-should.

McCarthy, John. "Review: The Question of Artificial Intelligence." http://jmc.stanford.edu/artificial-intelligence/reviews/bloomfield.pdf.

McNamara, Patrick. *The Neuroscience of Sleep and Dreams.* 2nd ed. Cambridge: Cambridge University Press, 2023.

Miller, Katharine. "When Algorithmic Fairness Fixes Fail: The Case for Keeping Humans in the Loop." Stanford University, November 2, 2020. https://hai.stanford.edu/news/when-algorithmic-fairness-fixes-fail-case-keeping-humans-loop.

Mitchell, Kevin J. *Free Agents: How Evolution Gave Us Free Will.* Princeton, NJ: Princeton University Press, 2023.

Morar, Florin S. "Reinventing Machines: The Transmission History of the Leibniz Calculator." *British Journal for the History of Science* 48, no. 1 (March 2015): 123–146.

Morris, Betsy. "Sharing the Technology's Weak Work Product Negatively Influences the Quality of Its Output Going Forward." *UCLA Anderson Review*, November 8, 2023. https://anderson-review.ucla.edu/ai-from-ai-a-future-of-generic-and-biased-online-content/.

Mosco, Vincent. *The Digital Sublime: Myth, Power, and Cyberspace.* Cambridge, MA: MIT Press, 2005.

Mullen, Brian, Craig Johnson, and Eduardo Salas. "Productivity Loss in Brainstorming Groups: A Meta-analytic Integration." *Basic and Applied Social Psychology* 12, no. 1 (March 1991): 3–23.

Napolitano, Christopher M. "More than Just a Simple Twist of Fate: Serendipitous Relations in Developmental Science." *Human Development* 56, no. 5 (2013): 291–318.

Noë, Alva. "Entanglement and Ecstasy in Dance, Music, and Philosophy." *Philosophy & Rhetoric* 54, no. 1 (2021): 63–80.

———. *The Entanglement: How Art and Philosophy Make Us What We Are.* Princeton: Princeton University Press, 2023.

———. *Varieties of Presence.* Cambridge, MA: Harvard University Press, 2012.

Noetel, Michael, Taren Sanders, Daniel Gallardo-Gómez, Paul Taylor, Borja del Pozo Cruz, Daniel Van Den Hoek, Jordan J. Smith, et al. "Effect of Exercise for Depression: Systematic Review and Network Meta-analysis of Randomised Controlled Trials." *BMJ* 384 (February 2024): https://www.bmj.com/content/384/bmj-2023-075847.

O'Gieblyn, Meghan. *God Human Animal Machine.* New York: Anchor Books, 2021.

Pacherie, Elisabeth, and Myrto Mylopoulos. "Beyond Automaticity: The Psychological Complexity of Skill." *Topoi* 40 (July 2021): 649–662.

Pacheco, Yesid J. O., and Virginia I. B. Toncel. "The Impact of School Closure on Children's Well-being during the COVID-19 Pandemic." *Asian Journal of Psychiatry* 67 (2022): 102957, https://www.ncbi.nlm.nih.gov/pmc/articles/PMC8641925/.

Pacherie, Elisabeth, "Toward a Dynamic Theory of Intentions," in *Does Consciousness*

Cause Behavior? Edited by Susan Pockett, William P. Banks, and Shaun Gallagher. Cambridge, MA: MIT Press, 2006.

Paries, Jean. "Lessons from the Hudson." In *Resilience Engineering in Practice*. Boca Raton, FL: CRC Press, 2017.

Parks, Rosa. *Rosa Parks: My Story*. New York: Puffin Books, 1999.

Pereboom, Derk. *Free Will.* Cambridge, UK: Cambridge University Press, 2022.

Pitcher, Laura. "Why Is AI Art So Cringe?" *Vice*, January 20, 2023. https://www.vice.com/en/article/m7gynq/why-is-ai-art-so-bad.

Porter, Michael E. "The Five Competitive Forces That Shape Strategy." *Harvard Business Review* 86, no. 1 (2008): 78–93.

Putnam, Robert D. *Bowling Alone: Revised and Updated: The Collapse and Revival of American Community*. New York: Simon & Schuster, 2020.

Ramseyer, Fabian T. "Non-verbal Synchrony in Psychotherapy: Embodiment at the Level of the Dyad." In *The Implications of Embodiment: Cognition and Communication*. Edited by Wolfgang Tschacher and Claudia Bergomi. Exeter, England: Imprint Academic, 2011.

Rens, Natalie, Gian Luca Lancia, Mattia Eluchans, Philipp Schwartenbeck, Ross Cunnington, and Giovanni Pezzulo. "Evidence for Entropy Maximisation in Human Free Choice Behaviour." *Cognition* 232 (March 2023): https://doi.org/10.1016/j.cognition.2022.105328.

Risko, Evan F., and Sam J. Gilbert. "Cognitive Offloading." *Trends in Cognitive Sciences* 20, no. 9 (September 2016): 676–688.

Rorty, Richard. *Contingency, Irony, and Solidarity*. Cambridge: Cambridge University Press, 1989.

Rowling, J. K. *Harry Potter and the Half-Blood Prince*. London: Bloomsbury, 2014.

Rushkoff, Douglas. *Survival of the Richest*. New York: Norton, 2022.

Salvaggio, Eryk. "The Hypothetical Image." *Cybernetic Forests*, October 29, 2023. https://www.cyberneticforests.com/news/social-diffusion-amp-the-I-of-the-digital-archive.

Schachter-Shalomi, Zalman. *The Geologist of the Soul: Talks on Rebbe-craft and Spiritual Leadership*. Boulder: Albion-Andalus Books, 2012.

Schaefer, Michael, Anne Reinhardt, Eileen Garbow, and Deborah Dressler. "Sweet Taste Experience Improves Prosocial Intentions and Attractiveness Ratings." *Psychological Research* 85 (June 2021): 1724–1731.

Schwab, Klaus. "The Fourth Industrial Revolution." *Rotman Management*, Fall 2016. https://www.rotman.utoronto.ca/Connect/Rotman-MAG/Issues/2016/Back-Issues---2016/Fall2016-TheDisruptiveIssue.

Simon, Herbert A. "Rational Choice and the Structure of the Environment." *Psychological Review* 63, no. 2 (1956): 129–138.

Smil, Vaclav. *Invention and Innovation: A Brief History of Hype and Failure*. Cambridge, MA: MIT Press, 2023.

Snyder, Timothy D. *The Road to Unfreedom: Russia, Europe, America.* New York: Crown Publishing, 2019.

Starling, Boris. "The End of Customer Service." *Perspective*, April 12, 2023. https://perspectivemag.co.uk/the-end-of-customer-service/.

Taylor, Astra. "The Automation Charade." *Logic(s)*, August 1, 2018. https://logicmag.io/failure/the-automation-charade/.

Thelen, Esther, Gregor Schoner, Christian Scheier, and Linda B. Smith. "The Dynamics of Embodiment: A Field Theory of Infant Perseverative Reaching." *Behavioral and Brain Sciences* 24, no. 1 (February 2001): 1–34.

Thomas, Rachel. "Medicine's Machine Learning Problem." *Boston Review*, January 4, 2021. https://www.bostonreview.net/articles/rachel-thomas-medicines-machine-learning-problem/.

Thomas, Wolfgang. "Algorithms: From Al-Khwarizmi to Turing and Beyond." In *Turing's Revolution: The Impact of His Ideas about Computability*. Edited by Giovanni Sommaruga and Thomas Strahm (Heidelberg, Germany: Springer, 2015): 29–42.

Thompson, Derek. "Why Americans Suddenly Stopped Hanging Out." *Atlantic*, February 14, 2024. https://www.theatlantic.com/ideas/archive/2024/02/america-decline-hanging-out/677451/.

Thurow, Lester C. "Needed: A New System of Intellectual Property Rights." *Harvard Business Review*, September–October 1997. https://hbr.org/1997/09/needed-a-new-system-of-intellectual-property-rights.

Torres, Émile P. "The Acronym Behind Our Wildest AI Dreams and Nightmares." *Truthdig*, June 15, 2023. https://www.truthdig.com/articles/the-acronym-behind-our-wildest-ai-dreams-and-nightmares/.

Tulchinsky, Igor, and Christopher E. Mason. *The Age of Prediction: Algorithms, AI, and the Shifting Shadows of Risk*. Cambridge, MA: MIT Press, 2023.

Turner, Fred. *From Counterculture to Cyberculture.* Chicago: University of Chicago Press, 2006.

Tyson, Courtney, Matthew J. Hornsey, and Fiona K. Barlow. "What Does It Mean to Feel Small? Three Dimensions of the Small Self." *Self and Identity* 21, no. 4 (2022): 387–405.

Verschure, Paul F., Cyriel M. Pennartz, and Giovanni Pezzulo. "The Why, What, Where, When and How of Goal-Directed Choice: Neuronal and Computational Principles." *Philosophical Transactions of the Royal Society of London. Series B, Biological Sciences* 369, no. 1655 (November 2014): https://doi.org/10.1098/rstb.2013.0483.

Vinsel, Lee. "Don't Get Distracted by the Hype around Generative AI." *MIT Sloan Management Review* 64, no. 3 (Spring 2023): 1–3.

Waddell, Kaveh. "Your Smart Devices Are Trying to Manipulate You with 'Deceptive Design.'" *Consumer Reports*, April 17, 2023. https://www.consumerreports.org/electronics/internet-of-things/smart-devices-trying-to-manipulate-you-with-dark-patterns-a6366326597/.

Wang, Long, Deepak Malhotra, and J. Keith Murnighan. "Economics Education and Greed." *Academy of Management Learning & Education* 10, no. 4 (2011): 643–660.

Wapner, Jessica. "Vision and Breathing May Be the Secrets to Surviving 2020." *Scientific American*, November 16, 2020. https://www.scientificamerican.com/article/vision-and-breathing-may-be-the-secrets-to-surviving-2020/.

Weil, Elizabeth. "Sam Altman Is the Oppenheimer of Our Age." *New York Magazine*, September 25, 2023. https://nymag.com/intelligencer/article/sam-altman-artificial-intelligence-openai-profile.html.

Weitzner, David. "Against 'Allyship.'" *Tablet*, November 4, 2021. https://www.tabletmag.com/sections/community/articles/against-allyship.

———. "Deconstruction Revisited: Implications of Theory over Methodology." *Journal of Management Inquiry* 16, no. 1 (March 2007): 43–54.

———. "No A.I. Is Smart Enough to Help Navigate Anxiety." *Psychology Today*, January 18, 2023. https://www.psychologytoday.com/us/blog/managing-with-meaning/202301/no-ai-is-smart-enough-to-help-navigate-anxiety.

———. "Three Ways Companies Are Getting Ethics Wrong." *MIT Sloan Management Review* 64, no. 1 (November 2022): 1–3.

———. *Connected Capitalism: How Jewish Wisdom Can Transform Work.* Toronto: University of Toronto Press, 2021.

———. *Fifteen Paths.* Toronto: ECW Press, 2019.

Weitzner, David, and James Darroch. "Fannie Mae." in *The SAGE Encyclopedia of Business Ethics and Society, Second Edition*, ed. Robert Kolb. Thousand Oaks, CA: Sage Publishing, 2018.

Wilde, Oscar. *The Picture of Dorian Grey.* New York: Random House, 2004.

Williams, Lawrence E., and John A. Bargh. "Experiencing Physical Warmth Promotes Interpersonal Warmth." *Science* 322, no. 5901 (October 2008): 606–607.

Wilson, Margaret. "Six Views of Embodied Cognition." *Psychonomic Bulletin & Review* 9, no. 4 (December 2002): 625–636.

Zhavoronkov, Alex. "Caution with AI-generated Content in Biomedicine." *Nature Medicine* 29, no. 3 (2023): 532, https://www.nature.com/articles/d41591-023-00014-w.

Zuboff, Soshana. *The Age of Surveillance Capitalism: The Fight for a Human Future at the New Frontier of Power*. New York: PublicAffairs, 2019.

Zweig, Katharina A. *Awkward Intelligence: Where AI Goes Wrong, Why It Matters, and What We Can Do about It*. Translated by Noah Harley. Cambridge, MA: MIT Press, 2022.

About the Author

photo © Stephanie Mills

David Weitzner is a writer, teacher, and consultant, working to uncover the best of what makes us human. He researches the science of artful thinking and creative co-creation, as well as strategies for bettering our ethical, spiritual, and business decision-making.

Dr. Weitzner is an associate professor of management at York University, with a PhD in Strategy, an MBA in Arts and Media Management and an Hon. BA in Philosophy. His paper *Harm Reduction, Solidarity, and Social Mobility as Target Functions* won the R. Edward Freeman Journal of Business Ethics Philosophy in Practice Best Paper Award, and his book *Connected Capitalism* won the Bronze INDIES Book of the Year Award for Business and Economics.

David writes the *Managing with Meaning* blog for *Psychology Today*, offering strategies for a more human-centric approach to business. His research has been published in prestigious peer-reviewed outlets such as the *Academy of Management Review, MIT Sloan Management Review, Organization Studies*, and *Journal of Business Ethics*. He also coedited *Corporate Social Responsibility* (Routledge) and coauthored *Strategic Management: Creating Competitive Advantages* (McGraw-Hill Ryerson).

David has presented as an invited guest at several high-profile international conferences, including Repurposing Management for the Public Good, hosted at the Møller Institute at the University of Cambridge, and Business as an Agent of World Benefit, co-sponsored by the UN Global Compact. His ideas have appeared in popular media outlets, including *NPR*, *Politico*, *Salon*, *The Globe and Mail*, *The Conversation*, *Business Insider*, *Tablet Magazine*, *The National Post*, *Yahoo News*, and *The Financial Post Business Magazine*, among others.

David lives in Toronto with his wife, three children, and a beagle. He continues to learn from artists, athletes, engineers, health care providers, religious clergy, and anyone else who operationalizes the mantra of "Think with BEAM and Defend your VICE."